HARRISON DECODED

HARRISON DECODED

Towards a Perfect Pendulum Clock

Edited by

Rory McEvoy and
Jonathan Betts

ROYAL
OBSERVATORY
GREENWICH

OXFORD
UNIVERSITY PRESS

OXFORD
UNIVERSITY PRESS

Great Clarendon Street, Oxford, OX2 6DP,
United Kingdom

Oxford University Press is a department of the University of Oxford.
It furthers the University's objective of excellence in research, scholarship,
and education by publishing worldwide. Oxford is a registered trade mark of
Oxford University Press in the UK and in certain other countries

Produced in association with Royal Museums Greenwich, the group name for the
National Maritime Museum, Royal Observatory Greenwich, the Queen's House and *Cutty Sark*.
www.rmg.co.uk

Published in the United States of America by Oxford University Press
198 Madison Avenue, New York, NY 10016, United States of America

British Library Cataloguing in Publication Data
Data available

Library of Congress Control Number: 2019947640

ISBN 978–0–19–881681–2

DOI: 10.1093/oso/9780198816812.003.0001

Printed and bound by
CPI Group (UK) Ltd, Croydon, CR0 4YY

Director's Introduction

It is now just over a quarter of a millennium since John Harrison's longitude watch, known as H4, was brought to the Royal Observatory for testing. Trialling the watch was the first of such tasks for the Astronomer Royal Nevil Maskelyne and marked the beginning of a new aspect to his role that enlarged exponentially. Soon after, Maskelyne was engaged in testing new experimental timekeepers by John Arnold, Thomas Mudge, and Thomas Earnshaw. It is pertinent to this project that he also tested a number of precision pendulum clocks, commissioned on behalf of prominent European astronomers. By 1800, the name of the Royal Observatory, Greenwich, was synonymous with expert time measurement.

With this history in mind, testing Martin Burgess's Clock B at the Royal Observatory offered a fantastic opportunity to reintroduce the practice of precision metrology at the observatory and add further interest and stimulation for visitors. This book offers a summary of the proceedings for three National Maritime Museum conferences that were convened at key stages of the testing process. In 2015, a formal peer-reviewed test of the clock was conducted with the assistance of the National Physical Laboratory, Worshipful Company of Clockmakers, British Horological Institute, and the Antiquarian Horological Society. The aim of the test was to see whether the clock could keep time to within one second in one hundred days. The clock's performance during the trial went above and beyond that of the best historic clocks used at Greenwich, earning the clock a Guinness World Record for 'The most accurate mechanical clock with a pendulum swinging in free air.'

However, it must be said that Clock B is no ordinary clock. It is the product of many years of collaborative study conducted by researchers, many of whom periodically met to discuss their ideas at the museum. This clock is not only a delightful piece of kinetic art but, moreover, it embodies an understanding formed through passionate research and has put Harrison's extraordinary theory into practice. We are grateful to Donald Saff for his acuity and drive to see this project come to fruition and his generous long-term loan of Clock B.

Dr Kevin Fewster AM
Former Director
Royal Museums Greenwich

Foreword

Throughout the forty-plus years John Harrison spent perfecting his successful solution to the longitude problem, he harboured another big idea for a fundamental improvement in precision timekeeping: he had devised an ultra-reliable pendulum regulator of novel design. The land-based equivalent to his marine timekeeper design, he asserted that this clock would be accurate to one second in one hundred days, and run continuously, without maintenance, for one hundred years.

But Harrison, busy building his sea clocks and defending them through repeated trials, never found the time or encouragement to finalise the regulator project. Since 1776, the year he died at age 83, his bold claims for the regulator's superior performance have echoed down the centuries, dismissed as empty boasts or even the ravings of a madman.

Harrison's ideal regulator has been realised at last, thanks largely to initial work begun by William Laycock and completed through the collaborative effort of a devoted research team determined to interpret Harrison's vision. The long journey from conception to completion of 'Clock B', capped by the instrument's rigorous testing at Greenwich, constitutes a tale to mirror Harrison's own travails.

Like H1 or H4, Clock B is a code word—a shorthand descriptor for a marvel of engineering. Martin Burgess's metallic, sculptural Clock B faithfully adheres to the principles Harrison set down in his writings, albeit in abstruse language that required 'decoding'. A product of patient experimentation, Clock B employs modern materials that were unavailable to Harrison, such as Invar, Teflon, and stainless steel. It traces its origins to the formation of the Harrison Research Group, first proposed by Mervyn Hobden in the mid-1970s and incorporating a group of experts who felt a bond with Harrison that transcended mere interest in his inventions. Thus there are several heroes in the story of how Clock B came to be, most of whom have their say in these pages. They all continue to regard (and frequently toast) John Harrison as 'the Master'.

Dava Sobel

Contents

List of Contributors

William Andrewes, who worked under the guidance of Martin Burgess between 1970 and 1972, is a museum consultant and sundial maker. He organized the Longitude Symposium and was editor of *The Quest for Longitude.*

Jonathan Betts is Curator Emeritus of Horology at the Royal Observatory, Greenwich. He is the author of several horological publications, including *Harrison* (1993), *Time Restored* (2006), and *Marine Chronometers at Greenwich* (2017) for which he carried out detailed studies on all of Harrison's timekeepers. He is Horological Adviser to The National Trust GB, The Harris (Belmont) Charity, and The Wallace Collection, and Curatorial Adviser to the Clockmakers' Company.

Martin Burgess trained as a silversmith and later worked with Egyptian antiquities before his pioneering work in sculptural horology and in research into the pendulum technology of John Harrison.

Martin Dorsch is a German-trained watchmaker, specialising in design and production technology at Charles Frodsham & Co.

David Harrison is Director of Technology Strategy at Ofcom. He holds a PhD on how nonlinearity affects oscillator performance.

M. K. Hobden is a former Herstmonceux chronometer section conservator at the Royal Greenwich Observatory. He specialises in time and frequency and quartz clocks.

Andrew King is a world-renowned horologist, an expert on John Harrison, and a clockmaker specializing in Harrisonian clocks.

Rory McEvoy, formerly Curator of Horology at the Royal Observatory, Greenwich, is now Lecturer in Horology at Birmingham City University.

Donald Saff is an artist, art historian, educator, and lecturer, specializing in American and English horology. He recently published *From Celestial to Terrestrial Timekeeping: Clockmaking in the Bond family,* through the Antiquarian Horological Society.

Dava Sobel is an award-winning former science reporter for the *New York Times* and writes frequently about science for several magazines, including *Audubon, Discover, Life,* and *Omni.* She is the author of international bestseller *Longitude* and *Galileo's Daughter.*

Roger Stevenson is head of workshop at Charles Frodsham & Co. He is a former conservator at the Royal Observatory, Greenwich and Herstmonceux chronometer section.

Tom Van Baak is an expert in digital measurement of timekeepers ranging from pendulum clocks to quartz and atomic frequency standards.

Philip Whyte has spent the past forty-five years involved in many aspects of horology and is one of the working directors of Charles Frodsham & Co.

1

Introducing the Precision Pendulum Clock

Rory McEvoy

Much of the collaborative study of John Harrison's unique pendulum clock system that informs many of the chapters in this volume has consisted of decoding the pertinent information contained within a curious publication written by Harrison in 1775. The full title of this work is *A Description Concerning Such Mechanism as Will Afford a Nice, or True Mensuration of Time; Together with Some Account of the Attempts for the Discovery of the Longitude by the Moon: as Also an Account of the Discovery of the Scale of Music* (Harrison, 1775; *CSM*).

The long-winded title does give the reader a real sense of what follows. This is an extraordinarily difficult read. There is a distinct economy of full-stops, footnotes occasionally span several pages, and there are even some footnotes with footnotes of their own. Furthermore, any clues as to how to go about constructing a Harrisonian pendulum clock are unordered and jumbled up among the extra subjects promised by the book's title. Throughout, the text is interspersed with anecdotal and often vitriolic accounts of his involvement in the longitude story, and in particular, his interactions with the 'professors (or priests)'. To further compound the ambiguity of the volume, *CSM* contains no illustrations of the clock nor of its components.

There is a deeper problem with the language of horology. There are a limited number of early treatises from which a reasonable glossary of terminology can be assembled. However, the practice of training and transfer of knowledge followed an oral tradition from master to apprentice and so numerous vocabularies have developed according to each master's lineage. Harrison, having arrived in the arena without formal training, had a vocabulary based on the texts that he had read and, in

McEvoy, R., *Introducing the Precision Pendulum Clock* In: *Harrison Decoded: Towards a Perfect Pendulum Clock*. Edited by Rory McEvoy and Jonathan Betts, Oxford University Press (2020). © Oxford University Press. DOI: 10.1093/oso/9780198816812.003.0001 connection to his own work, developed new terminologies that were not readily translatable. For example, Harrison uses the word 'dominion' to describe the dynamic relationship between the pendulum and the escapement. This single word encompasses a concept that is described later on in Chapter 9. This type of communication problem is comparable, perhaps, to words such as the Danish word *hygge*, which conveys a feeling of contentment that is bound to a unique cultural paradigm, and therefore cannot be translated to a single word from the English language.

A contemporary review of *CSM* opens thus:

> The curiosity of the Public may perhaps be raised in expectation of having the principles of Mr Harrison's celebrated time-keeper fully explained, the many curious contrivances in his machine clearly described, and their uses pointed out by the inventor himself. We are sorry to say the Public will be disappointed. (Anon, 1776:330)

It is hoped that this volume will redress the balance and lay bare the intricacies of Harrison's pendulum clock system. This chapter will provide the reader with an overview of the mainstream development of the precision pendulum clock, which is intended to provide some historical context to Harrison's work and complement the other chapters in this volume by explaining some of the key physical problems that affect mechanical pendulum clocks.

The pendulum as a time-measuring device

The pendulum's history as a timekeeping device begins around 1600. Galileo Galilei, as perpetuated by his student and first biographer, famously observed the isochronous properties of a swinging lamp in Pisa Cathedral. Using his pulse as a time standard he concluded that the larger arcs of swing took the same time to complete their action as the shorter ones. Galileo went on to use the pendulum as a time standard in his experiments, intending to derive a constant for natural acceleration. However, the pendulum alone was not an appropriate instrument with which to time the descent of rolling balls down an inclined plane, for example. There was too much room for error in observing the pendulum by eye in conjunction with a fast-moving sphere.

The pendulum served to calibrate a simple form of water clock, known as an outflow clepsydra. This device consisted of a large water-filled vessel with a small opening at the bottom that allowed a constant stream of water to escape. Galileo collected water from the clepsydra in a glass. The amount of water collected was metered by the oscillations of the pendulum. In this manner, Galileo was able to provide an arbitrary time measurement in terms of the weight of water collected. This method provided relative values for distance travelled over a given period. In *Dialogo,* he presents his results for acceleration in this manner—distance over weight of water collected (Crew and Salvio, 1914:179).

The pendulum, in its simplest form, is a weight (the bob) suspended from a fixed point, which when set in motion swings from side to side. A left-to-right motion of the bob is called a vibration and a full cycle, say, left to right and back again, is known as an oscillation. By experimenting with bobs made from both lead and cork, Galileo was able to conclude that the weight or density of the pendulum bob did not affect its period, though air resistance reduced the cork bob's arc of swing much faster than its lead counterpart. Further experiments demonstrated that it was the length of the suspension that dictated the duration (period) of a vibration. He also found that the relationship between length and period was not linear but was governed by a square law. For example, to double the period of a pendulum, the length of the suspension needs to be multiplied by four.

Presenting time in terms of weight of water collected was not ideal. Galileo, assisted by four 'patient and curious friends', attempted to quantify the period of the pendulum. The arduous process of maintaining and counting the vibrations of a simple pendulum was described in a letter to his friend and long-term correspondent, Giovani Battista Balliani (1582–1666). Between successive transits of a bright star, the team counted a total of 234,567 vibrations during the 24-hour period, which suggests that their pendulum had a period of around one-third of a second. However, the consecutive nature of the digits in his total number of vibrations and the fact that he continued to use weight rather than seconds of time in *Dialogo,* published after the experiment, suggests that he was not confident in the result (Drake, 1978:399).

Other philosophers, such as Marin Mersenne (1588–1648) and Giambattista Riccioli (1598–1671), also attempted to identify the length of a pendulum with a one-second vibration in a similar fashion. Again, the 24-hour manual process proved too laborious and attempts to shorten the time taken to calibrate the pendulum by use of other instruments, such as clocks, sand glasses and sundials, were unsatisfactory due to their imprecision. An automated system that could automatically maintain and count the pendulum's vibrations was required—i.e. a mechanical pendulum clock. The first published description of such a device came from the brilliant Dutch mathematician and astronomer Christiaan Huygens (1629–1695) in the 1658 pamphlet *Horologium* (Edwardes, 1970:35).

Circular deviation

By this time, it was understood that the pendulum was not isochronous, as Galileo had assumed. If a pendulum follows a circular arc it will inevitably take longer to swing over a large arc than a short one. This characteristic is commonly known as circular deviation. The French theologian and philosopher Mersenne was aware of this but, like Galileo, he did not factor it into his pendulum experiments. Huygens identified that a pendulum swinging along a cycloidal path would have the same period, regardless of its amplitude. The cycloidal curve is drawn by a single point on the circumference of a circle that rolls along a straight line. Importantly in pendulums, the cycloidal path means that the pendulum's effective length shortens as the arc increases, and so with the correct curve the pendulum becomes isochronous.

In the fourth section of *Horologium Oscillatorium* (1673), Huygens credits Mersenne for introducing him to the cycloid. The two men never met but corresponded after an introduction by Christiaan's father, Constantijn Huygens (1596–1687). Huygens created a desirable path for the pendulum by means of a pair of metal curved cheeks that enclosed the pendulum's silk suspension and, in theory, they rendered the pendulum isochronous by shortening its effective length as the arcs of swing widened. With this device, Huygens's mechanical pendulum clock was substantially better than its best-performing predecessors, bringing the daily instability down from around one minute to around ten seconds (Yoder, 2004:12).

Unpredictable changes in the pendulum's arc of swing are unavoidable in a mechanical clock of this early type. The following factors are listed roughly in order of their effect on such a clock's timekeeping: variations in the energy transferred to the pendulum by the escapement, caused by mechanical imperfections in the clock movement; the changing properties or location of the oil lubricating the clock's moving parts; and changes in barometric pressure, and particularly temperature. It is important to note also that the effect of circular deviation is not linear. The relative change in period increases in the larger arcs of swing. For this reason, clockmakers attempted to minimise the pendulum's arc of swing to diminish the effect of circular error.

Towards a perfect oscillator

In scientific terms, clock pendulums are oscillators. Depending on the way a pendulum is used or maintained, it can be categorized in different ways. For example, Huygens's cycloidal cheeks is close in principle to being a linear oscillator—where the speed of the return of the pendulum increases with an increase in amplitude (or vice versa) and thereby maintains a constant period. However, in practice the mechanical movements that drove the pendulums had too much of an influence over the pendulum's motion for the system to be effective or come close this theoretical ideal.

The Curator of Experiments at the Royal Society in London, Robert Hooke (1635–1703), experimented with pendulum clock design, pursuing a different arrangement, where a heavy pendulum bob received a small impulse and minimised circular deviation by swinging over a very small arc. In 1669, Hooke demonstrated his design to the Royal Society, using a 3-lb lead ball suspended on a string of around 14 feet in length and an adapted a pocket watch movement to keep the pendulum swinging, making one vibration in two seconds. In theoretical terms, Hooke's pendulum was a simple harmonic oscillator and arguably set the template for the precision pendulum clocks through to the late nineteenth century.

The subsequent discussions that ensued at the Royal Society following Hooke's demonstration of the design are scantly reported, but are fundamental not only to the history of the precision pendulum clock but also the history of the Royal Observatory, Greenwich. Hooke's remarks on the design are reported in the journals of the Royal Society and, on 26 June 1669, he made these two important statements regarding his design: 'the smallness of the vibrations renders the pendulum insensible of the impression, which the watch makes upon it, said, that the weight appendant to the string was so great, that the impression could have no power upon it' (Birch, 1756:388). Firstly, the impulse given by the watch was so small in proportion to the energy stored in the swinging bob that it could maintain but have almost no effect on the amplitude or period. Secondly, the low-energy impulse from the pocket watch movement was delivered to the bob and, therefore, was unable to distort the pendulum's shape by curving the string. The ill-effects of energy loss through distortion of the pendulum's shape evidently concerned clockmakers. Some years later, William Derham (1657–1735), author of one of the first clockmaking treatises in the English language, described how one of his clock pendulums was con-structed specifically to resist deformation: 'the pendulum rod flat & strong, broad at the bottom, & tapering all the way to the top. But without such a provision, the rod by bending…makes considerable alterations in the length of the vibrations' (Derham, 1714).

In 1675 the Royal Observatory was founded and John Flamsteed (1646–1719), the first Astronomer Royal, was provided with two excep-tional pendulum clocks that were integral to the structure of the room, known today as the Octagon Room. These two clocks were made by the famous clock- and watchmaker Thomas Tompion (1639–1713) and were designed to run for one year between windings. Each had a two-second pendulum suspended above its movement and, as in Hooke's 1669 demonstration, they maintained their pendulum's swing from below, with a very light touch. When the Observatory's architect Christopher Wren (1632–1723) saw Hooke's timekeeper in 1669 he sug-gested that it could be improved by use of a 'cylindrical staff of 28 feet long, and making it move in the middle on a pin, and hanging an equal weight on each end of it, to be moved with a pocket watch.' Had this idea been implemented at Greenwich, the shape of the Octagon Room would have necessarily been vastly different to accommodate such a clock (Birch, 1756:361).

The clocks were indeed far superior to anything else of the time in terms of their timekeeping; however, they were very problematic and stopped regularly, requiring cleaning and reoiling, across the early years of their use (Howse, 1970:27). Despite the frequent stoppages, the clocks enabled Flamsteed to determine that the Earth's speed of rota-tion was constant throughout the year. Previously, this was assumed and Flamsteed's assertion became a solid foundation for the positional astronomy at Greenwich that followed. However, in the late nine-teenth century, inconsistency between predicted and observed posi-tions of the Moon revealed to astronomers that the speed of the Earth's rotation showed some seasonal fluctuation. This seasonal disparity was far smaller than the daily instability of Flamsteed's clocks at Greenwich. Flamsteed's records of the clocks' performance in 1677 show that they

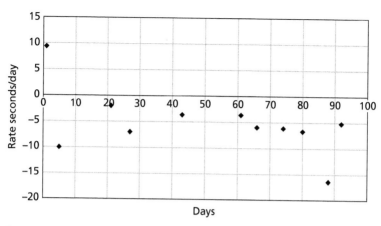

Fig. 1.1 Average daily rate of a Tompion year-going clock (with 'pivoted pendulum') at Greenwich from 16 March to 19 May 1677. © Rory McEvoy

could be relied on to keep time to within around six seconds per day (Fig. 1.1).

The deadbeat escapement

Flamsteed corresponded regularly with Richard Towneley (1629–1707), a friend of Flamsteed's patron Sir Jonas Moore (1617–1679), keeping him up to date on developments in London. From surviving letters we have learned that Towneley was likely to have been the first to make a form of the deadbeat escapement and that Thomas Tompion (1636–1713) made a clock for Moore with a similar escapement soon after. Flamsteed wrote to Towneley in September 1675: 'I hear not of any pallets for pendulums that have been made your way, but Mr Tompion likes it very well since as the other it puts not the second finger back by girds.'

The term 'gird' refers to the action of the second hand when attached to a recoiling escape wheel, which is never stationary: as it advances one division (two seconds on Flamsteed's clocks), it moves backward (recoils) before advancing to the next (Howse, 1970:18–34). It is interesting to speculate as to what exactly appealed to Tompion; perhaps it was the precise motion of a deadbeating second hand that enabled a clear reading to the nearest second, or maybe it was a mechanical advantage that encouraged isochronism in the pendulum.

Flamsteed wrote to Towneley, while observing from the Queen's House at Greenwich, and described the poor performance of his pendulum clock. He explained that the movement had become clogged with dirt and that the low energy imparted to the pendulum had caused the clock's timekeeping to shift from a loss of thirty seconds per day to a gain of one minute per day. Flamsteed informed Towneley that the escapement was of the 'old form', which was likely to have been an anchor escapement with a recoiling action (Howse, 1970:23). The reported change in timekeeping suggests that initially the losing rate was caused by the distinctive properties of the anchor escapement. Such an escapement will cause a slowing of the rate if the power is reduced. This dramatic increase in clock rate was likely caused by some mechanical failure, as it simply is too large a shift to be caused by circular deviation alone. In the same letter, Flamsteed intimated that Moore was of the opinion that the deadbeat escapement might have helped to isochronise the clock.

Escapement error

This term is commonly used to describe the characteristics of a particular escapement. The term is a little misleading as 'error' implies a negative quality. In the simplest of terms, this describes the effect caused by the presence of the escapement on the period of the oscillator and the resultant alteration to clock rate at different pendulum amplitudes. The English mathematician and surveyor Charles Hutton (1737–1823) chronicled that George Graham (c.1673–1751) and Edmond Halley (1656–1742) conducted a series of experiments at the Royal Observatory and they had concluded that it was the recoiling action of the escapement that caused the acceleration in rate as the arc of the pendulum's swing increased. Hutton also identified Graham as the inventor of the deadbeat escapement. In doing so, he stated that Graham 'restored to the pendulum wholly in theory, and nearly in practice, all its natural properties in it is detached state' (Hutton, 1795:419).

However, in practice the deadbeat escapement has its own unique characteristics, and variations in the pendulum's amplitude also cause changes in rate. Experience shows that the presence of the deadbeat escapement slows the rate of the clock and does so increasingly as the driving force increases. Hutton's piece bolstered the misunderstanding that Graham was the inventor, and indeed the fact that the deadbeat was a superior design. If one analyses the output from Graham's business over its thirty-eight-year lifespan, and considers the strict maintenance of a house style, established by Tompion, it is difficult to maintain the assertion that the deadbeat escapement is the superior design. During Graham's tenure of the business, around 2,500 timepiece pocket watches were produced (based on serial numbering of extant watches). In the mid-1720s, Graham abandoned the use of the verge in favour of the cylinder escapement in his watches. From the subsequent production of around 1,500 watches, only one surviving example features the verge escapement (number 5999). This precedent raises an important question—why weren't all of Graham's pendulum clocks fitted with the superior deadbeat escapement? Graham's longcase clocks feature both forms of escapement and there is no evidence to suggest a preference.

Another eighteenth-century exponent of the deadbeat escapement was Alexander Cumming (1733–1814), who wrote:

> That the influence of the oil and friction, is always less on the dead-beat, than on the recoil; all other circumstances being alike...the recoil can have no tendency to keep the vibrations of more equal length. Therefore, that in all cases whatsoever, the DEAD-BEAT is preferable to the RECOIL. (Cumming, 1766)

The Reverend William Ludlam (1716–1788) was not as dogmatic on the subject of clock escapements and gave a more open-minded view of contemporary attempts to improve clockwork, including good analysis of the work of John Harrison (1693–1775) and other recoiling escapements. Ludlam's occasional references to Cumming's publication are somewhat scathing. He suggested that Cumming had 'grossly misunderstood' some of Harrison's ideas and that his suggestion of suspending the pendulum from a substantial block of marble set into the wall was the best piece of advice in the whole book. Ludlam (1769:138) concluded:

> How far these inventions may improve clocks remains to be tried. In the mean time, if the pendulum be properly suspended, and its length not subject to be changed by the weather; a clock of the common construction with dead seconds will go well enough for any astronomical purposes whatever...the observatory clock could always be depended upon for ten days or a fortnight, so as not to gain or lose in that time above a second or two at the most, an astronomer must have bad luck indeed, if

in that time he cannot take a [*sic*] observation either of the Sun or stars by which he may examine the going of his clock, and determine its error.

Weather and the pendulum

Ludlam's description of the typical astronomical clock mentions the importance of having compensation for 'weather' on the pendulum. To fully understand the Harrison system, it is essential to have a working knowledge of how changes in environmental conditions affect pendulum clocks. Harrison deliberately made his pendulum more susceptible to changing conditions of the air to enable compensation— unlike the majority of precision clockmakers who opted to avoid or, at best, minimise the effects of changing conditions.

From the early days of the pendulum clock, natural philosophers were keen to interrogate the physical properties of the pendulum and the vacuum pump was a valuable tool in such endeavours. Robert Boyle (1627–1691) conducted early experiments with free-swinging pendulums inside an evacuated chamber. Boyle's goal was to study the effects of air pressure on the decay in amplitude of an undriven a pendulum and, from his experiments, determined that there was 'no sensible difference between the celerity of a pendulum's motion in the air and that *in vacuo* [sic]' (Birch, 1756:429).

In 1704, William Derham (1657–1735) also studied the pendulum in an evacuated chamber. Unlike Boyle, he placed a pendulum clock in the receiver to study the effects of air pressure on timekeeping. He observed that the pendulum's arc of swing increased as the air was evacuated and that the clock slowed by two seconds per hour when the vibrations were at their largest. This was an anticipated product of circular deviation. However, Derham suspected that the pendulum was moving faster in the vacuum and through 'nice experiments' was able to confirm his suspicion. He demonstrated this by running the clock in air and adding to the driving weight until the same enlarged amplitude was reached. The clock ran almost three times slower than it had in the vacuum.

Derham's experiments very neatly illustrate the following effects of changing air density on ordinary pendulum clocks. A reduction in air density will reduce air resistance and thereby increase the amplitude of the pendulum. Circular deviation will cause the period to increase; however, the thinner air offers less buoyancy to the pendulum, and so the gravitational pull on it is greater, which causes the bob to travel at a greater velocity and reduces the effect of circular deviation. Derham's latter experiment showed that an increase in air density will have the opposite effect (Derham, 1735).

The shifting levels of gravity and air resistance acting on the pendulum can be caused by both barometric pressure and temperature. However, these two causes do not have the same effect. On the one hand, a change in barometric pressure alone will affect a pendulum clock in the same manner as described in Derham's experiments with the evacuated receiver. On the other hand, colder temperatures will increase air density but the secondary effect of drag, the air's viscosity, differs. Despite the increased density, the colder air is more inert and so the pendulum bob experiences less drag. Temperature also has a greater influence on clock rate, as the materials that make up the clock movement—particularly the pendulum—expand and contract in changes of temperature.

Whilst George Graham never claimed priority of invention for the deadbeat or cylinder escapements, he did invent the first temperature-compensated pendulum (Graham, 1722). Because the period of a pendulum is governed by its length and because most pendulums were made from brass, steel, and lead, when the temperature rises, the pendulum's components expand and thereby slow the clock's rate. By replacing the pendulum bob with a mercury-filled glass jar, Graham's pendulum maintains a constant centre of gravity in changing temperatures. The upward expansion of the mercury in the jar compensated for the elongation of the metal rod. The consequence of changing air density due to temperature variation is often unconsciously compensated for by the clockmaker. In a mercurial pendulum, the compensation is achieved by fine-tuning the amount of mercury in the jar.

To improve or not to improve

In 1749, a clock with all of the refinements mentioned previously was purchased for use at the Royal Observatory by the third Astronomer Royal, James Bradley (1693–1762). The clock, known today as Graham No. 3, was the epitome of an excellent astronomical clock with deadbeat escapement and temperature-compensated pendulum (Harrison's gridiron)—the type that Ludlam advocates 'will go well enough for any astronomical purposes whatever.' The integrity of Ludlam's statement is upheld by the fact that the format of the majority of such clocks remained mostly unchanged throughout the history of mechanical clock production. However, Graham No. 3 contains rich archaeology that illustrates repeated attempts to improve its performance. Its history bears testament to the fact that Nevil Maskelyne (1732–1811) did not consider the design to be adequate and throughout his tenure at Greenwich continually sought a better time standard for the Observatory.

The success of Harrison's fourth timekeeper opened up a new facet to the role of Astronomer Royal. Maskelyne, being the country's foremost expert in time derivation, was charged with testing the nascent marine timekeeping technology. The first trial at Greenwich was that of Harrison's fourth timekeeper, known today as H4, and the daily record of the watch's timekeeping was published alongside the error of Graham No. 3, temperature and barometer readings.

Astronomers at the Royal Observatory determined the daily error of the clock by observation of the transits of bright stars across what is known today as the Bradley Meridian. These stars were often referred to as clock stars for they were used to derive local sidereal time. To make an observation, they read the time to the nearest second from the clock before turning their eye to the telescope. Then, continuing to count the audible ticking of the clock, they observed the relative positions of the star, for two successive ticks, either side of the vertical wires that segmented the field of view from the telescope. They used a graticule in the eyepiece to gather a series of timings from which an average could be made to obtain the most precise value for the clock's error at the time of the transit.

The published record of the trial is an extraordinary document that encapsulates the day-to-day performance of one of the country's best clocks in the mid-eighteenth century (Maskelyne, 1767). Figure 1.2 is drawn using Maskelyne's published results and follows his method of presentation. Positive values indicate that sidereal time was ahead of the clock, therefore that the clock was losing, and vice versa. Additionally, the average daily temperature and air pressure were recorded; these values are represented on the chart above and below the clock rate for ease of reading. From this data one can infer that the clock was, on the whole, a reliable timekeeper that kept time to around a tenth of a second for most days, albeit with a gaining or losing rate. However, there are numerous examples where the clock's error changed by as much as

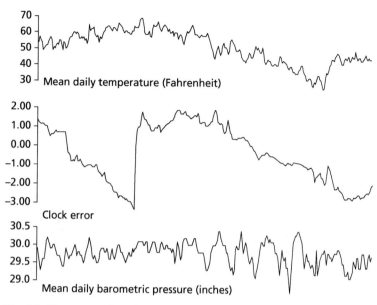

Fig. 1.2 The going of Graham No. 3 during the trial of H4 at the Royal Observatory between 6 May 1766 and 4 March 1767. © Rory McEvoy

half a second in one day. Three months into the trial of H4, Maskelyne's assistant, Joseph Dymond, stopped the clock and adjusted the pendulum length to correct its gaining rate. This event caused the sudden shift in the graph, seen one-third of the way along (Maskelyne, 1767:xii).

The data indicate an occasional correlation between clock rate and changes in barometric pressure, though not with great regularity. There are some highs and lows in barometric pressure that do not appear to have any discernible effect on the clock's rate. Additionally, there are shifts in the rate that appear unrelated to changes in environmental conditions and which are likely to have been the result of shifting states of lubrication within the clock movement. Dymond's intervention in early August follows Ludlam's summary of the use of pendulum clocks in the Observatory—they were somewhat wayward servants and required regular correction according to the astronomers' observations.

In October 1794, Maskelyne's assistant David Kinnebrooke Jnr described to his father the instruments that he used at Greenwich, and in his (1794) letter he remarked that Harrison, Arnold, Kendall, and

Earnshaw had all had a hand in improving Graham No. 3. Beyond Kinnebrooke's letter, surviving records do not confirm Harrison's involvement with Graham No. 3. It is possible that he was confusing some of the upgrades to the clock, such as incorporation of Harrison's design for maintaining power and the mounting of the pendulum directly to the stone pier, as the work of Harrison. Evidently, Harrison had lost all interest in helping the Astronomer Royal to improve his instrumentation and did not offer a pendulum clock of his design to Greenwich. In his final written work he acerbically remarked, 'I once thought of giving a clock to the Observatory at Greenwich, but my bad usage proved too tedious for that' (Harrison, 1775:52).

In the years following the Observatory trial of H4, Graham No. 3 was subjected to a series of improvements, made by the precision watch-makers on Kinnebrooke's list. It is not a coincidence that the timing of these interventions coincided with the deliveries of new marine timekeepers for testing at the Observatory (McEvoy, 2014). Maskelyne's observations fed into calculations that provided the predicted relative positions of the planets, stars, and the Moon in the *Nautical Almanac*. Inaccuracy could result in loss of life at sea and so it is unsurprising that Maskelyne actively sought to improve his instrumentation. Indeed, in the late nineteenth century, thanks to improved observation equipment, astronomers noticed a regular disparity between observed and predicted positions of the Moon. This disparity was later identified as a product of seasonal fluctuations in the Earth's speed of rotation.

In 1807 Maskelyne took delivery of a new design of transit clock from William Hardy (d.1832) for trial at the Observatory. Maskelyne records the clock as having a detached escapement, suggesting that the pendulum was free of unwanted frictional influences. In fact, the clock had a spring pallet escapement, which can more correctly be categorized as a constant force escapement. The trial showed that the design offered far greater accuracy than Graham No. 3. Maskelyne's (1807) manuscript record has been represented graphically using a scatter plot (Fig. 1.3). This type of chart effectively illustrates the clock's margins for error. From day 20 onward, the clock consistently indicated time to within half a second. However, the design was not without its problems. During the first twenty days, the clock's performance was erratic because of a fundamental flaw in the design. During this period, the lubrication of the escapement had failed due to ingress of dirt. The clock's stability was absolutely dependant on the condition of the

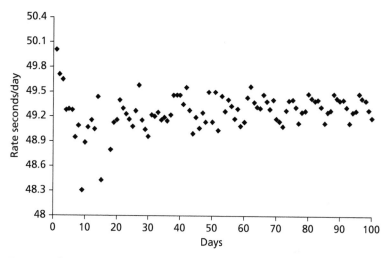

Fig. 1.3 The going of William Hardy's clock at the Royal Observatory from 29 June to 13 July 1807. © Rory McEvoy

escapement's lubrication. Maskelyne noted that on days 17 and 18, the clock pallets were cleaned and reoiled.

The good performance of Hardy's clock was partly due to the fine and careful construction of the wheelwork. Due to the escapement's dependency on a perfect state of lubrication, the new transit clock was converted to deadbeat escapement by the ambitious E. J. Dent (1790–1853), not long after its installation at Greenwich (Plate 1).

The nineteenth century saw continued development in pendulum clockmaking, and some notable examples were of a sufficient quality to be predictably affected by changes in barometric pressure. Dent no. 1906, for example, served as the Sidereal Standard at Greenwich for over forty years. When first installed in the magnetic basement in 1871, its timekeeping was so stable that it responded predictably to changes in air pressure; a rise of one inch of mercury caused the clock to lose 0.3 seconds per day (Airy, 1872:xxiv) Today, this clock has a glazed door, fitted for display at the Festival of Britain in 1951, to show off the barometric compensation. A mercury-filled J-tube (typical of the type found in wheel barometers) and a see-saw type arrangement with a float at one end and a horseshoe magnet at the other corrected the period of the pendulum, compensating for changing air density by more or less interaction with two bar magnets strapped to the pendulum bob.

Thomas Romney Robinson (1792–1882), the astronomer at Armagh Observatory, had previously attempted to compensate for changes in air density by attaching a barometer tube to the clock pendulum. In his 1843 paper, he mentioned Edward Sabine's (1788–1883) pendulum experiments that had been conducted inside an evacuated chamber and expressed a wish that a transit clock could be made to run inside an evacuated chamber (Robinson, 1843:18). Edmund Beckett Denison (1816–1905) advocated the use of a large arc of swing in larger clocks to use circular deviation to correct for the effects of changing air density (Beckett, 1903:74). This fine-balancing act is challenging enough, but with frictional escapements, at the large amplitude required, they may produce greater errors than the air density issue that was being corrected for (Robertson, 1929:196).

Towards the end of the nineteenth century, Robinson's wish for a clock that could be run reliably in a tank of constant pressure became a reality. The first commercial clocks were produced by Sigmund Riefler. His clocks were electrically rewound every thirty seconds and featured a form of deadbeating escapement that was connected directly to the pendulum through the suspension. According to the company's marketing material, the clock did offer tremendous stability as a timekeeper of around one hundredth of a second per day. In the early 1920s, a railway engineer, William Hamilton Shortt, devised a way to maintain a pendulum in an evacuated chamber without it being directly connected to an escapement, thus alleviating the clock from the associated problems of escapement error and inconsistency.

Shortt's free pendulum came close to meeting the desired conditions of a stable harmonic oscillator (Plate 2). The pendulum bob had a high stored energy, relative to the loss incurred during oscillation, and only required a minimal impulse every thirty seconds to maintain an almost constant arc of swing. Physicists describe this pendulum as having a high Quality (Q) factor. The Shortt system used Invar, an iron and nickel alloy with a very low coefficient of thermal expansion, for the pendulum rod and so it should have been free from temperature error. However, staff at the Royal Observatory found that the clocks did respond to changes in temperature, and also that the amplitude of the pendulums varied, albeit slightly (Jackson and Bowyer, 1928:480).

Staff at Greenwich observed temperature change and the pendulum's amplitude (in millimetres) to make mathematical corrections to the rate. To minimise the effect of circular deviation, Shortt Free

Pendulums operated at a running arc of just under two degrees. However, even with this low arc of swing, the effect of a miniscule change in amplitude was quantifiable. A change of just one hundredth of a millimetre to the semi-arc would amount to an error of a half-second after one year (Hope-Jones, 1930:157). Once the clock rate had been smoothed, there were still inconsistencies evident in the time-keeping.

Changes in gravity, caused by the Sun and Moon, affected the timekeeping. This effect was likewise mathematically corrected. Invar showed some material instability and this was identified by the varying performance between Shortt clocks. In addition to these, there was one unpredictable influence: seismic activity. It is arguable that this influence is one of the worst enemies of stability in high-Q pendulum systems. An extreme example of seismic interference was captured at the Lick Observatory, California by a photographic arc recorder that monitored their Shortt Free Pendulum. The quake caused a large increase in amplitude, and therefore a slowing of the clock's rate. Because of the pendulum's high-Q factor, it took over 24 hours for the amplitude to return to the running arc (Jeffers, 1935:79).

The fact that noise vibrations affect clock pendulums, and in an urban environment they come from a multitude of unpredictable sources, is nothing new to clockmakers. In the mid-1730s James Bradley (1693–1762) described how a pendulum clock, used in a gravity experiment, was deliberately set up 'in a room, situated backward from the street, and on the north side of his [George Graham's] house, to prevent its being disturbed by coaches, or other carriages that passed through the streets' (Bradley, 1734). To demonstrate this issue in a modern context, a short film of the pendulum of the museum's Fedchenko tank regulator was shown at the first conference on Clock B at the National Maritime Museum in June 2014. The clock, displayed in the Time and Greenwich gallery at the Royal Observatory, had no electrical supply at the time and yet the pendulum was oscillating with amplitude that was certainly visible to the naked eye.

This history has remained within the walls of the Royal Observatory, Greenwich and for this reason will draw to a close at the Shortt Free Pendulum Clock. Its mechanical DNA descended from Hooke's design and the first clocks supplied to John Flamsteed at the Observatory's foundation: a large and heavy oscillating weight with light impulse that maintained its swing over a small arc. The Shortt system approached perfection but only with the assistance of mathematical smoothing of its time indication. It achieved a level of precision that revealed the minutest of problems and yet, by virtue of the fact that it required anchoring to the Earth, it was always going to be influenced by noise vibration.

So, perhaps 250 years too late, Martin Burgess's Clock B may have demonstrated that Harrison was on to something remarkable and that there was another way to make an accurate pendulum clock. The ensuing chapters will chart this extraordinary history, reveal the dichotomy between the Hooke and Harrison methods, report on the modern-day practical investigation, and present the perceived theoretical approach that almost died with its architect.

References

Airy, G. (1872). *Observations made at the Royal Observatory in the year 1872.* London: Eyre and Spottiswood.

Anon. (1776). *The Monthly Review or, Literary Journal: from July 1775, to January 1776.* Vol. 43.

Beckett, E. (1903). *A Rudimentary Treatise on Clocks, Watches, and Bells for Public Purposes.* London: Crosby, Lockwood and Son.

Birch, T. (1756). *The History of the Royal Society of London,* Volume II. London: Royal Society.

Bradley, J. (1734). Observations made in London, by Mr George Graham; and at Black-river Jamaica, by Mr Colin Campbell, about the going of a clock; in order to determine the difference between the lengthes of isochronal pendulums in those places. *Philosophical Transactions of the Royal Society.* London: Royal Society.

Crew, H. and Salvio, A. (1914). *Dialogues Concerning Two New Sciences, Galileo Galilei.* New York: Dover.

Cumming, A. (1766). *The Elements of Clock and watch-work.* London.

Derham, W. (1714). Observations concerning the motion of Chronometers. Royal Society ref. Cl.P/3ii/10/.

Derham, W. (1735). Experiments concerning the vibrations of pendulums. *Philosophical Transactions of the Royal Society.* London: Royal Society.

Drake, S. (1978). *Gallileo at Work: His Scientific Biography.* Chicago: University of Chicago Press.

Edwardes, E. (1970). Horologium by Christiaan Huygens, 1658: an English translation together with the original Latin text in facsimile. *Antiquarian Horology* 7(1), 35–55.

Graham, G. (1722). A contrivance to avoid the irregularities in clock's motion, occasion'd by the action of heat and cold upon the rod of the pendulum. *Philosophical Transactions of the Royal Society*. London: Royal Society.

Harrison, J. (1775). *A Description Concerning Such Mechanism*. London, Jones, T. https://ahsoc.contentfiles.net/media/assets/file/Concerning_Such_Mechanism.pdf [accessed 12 December 2018].

Hope-Jones, F. (1930). The Shortt Free Pendulum Clocks at Greenwich. *Horological Journal*. April, 157–158.

Howse, D. (1970). The Tompion Clocks at Greenwich and the Dead-Beat Escapement. *Antiquarian Horology*, December, 18–34.

Hutton, C. (1795). *Mathematical and philosophical dictionary*, vol. II. London: Johnson, J.

Jackson, J. and Bowyer, W. (1928). The Shortt Clocks at the Royal Observatory, Greenwich. *Proceedings of the RAS*, March.

Jeffers, H. M. (1935). The Shortt Clock of the Lick Observatory. *Lick Observatory Bulletin*, no. 468.

Kinnebrooke, D. (15 October 1794). Photocopied letter. Cambridge University Library, RGO 35/98.

Ludlam, W. (1769). *Astronomical observations made In St. John's College, Cambridge, in the years 1767 and 1768: With an account of several astronomical instruments*. Cambridge: Archdeacon, J.

Maskelyne, N. (1767). *An account of the going of Mr John Harrison's watch, at the Royal Observatory, from May 6th, 1766, to March 4th, 1767*. London: Richardson and Clarke.

Maskelyne, N. (1807). Account of the going of Mr William Hardy's clock with detached escapement at the Royal Observatory. Cambridge University Library, RGO Vol. 1, p. 243.

McEvoy, R. (2014). Maskelyne's time. In Higgett, R. (ed.) *Maskelyne, Astronomer Royal*, pp. 168–193. London: Robert Hale.

Robertson, D. (1929). The theory of pendulums and escapements. *Horological Journal*, March, 194–196.

Robinson, T. R. (1843). *On the barometric compensation of the pendulum*. Report of the thirteenth meeting of the British Association for the Advancement of Science. London: John Murray, Albemarle St.

Yoder, J. G. (2004). *Unrolling Time: Christiaan Huygens and the Mathematization of Nature*, Cambridge, UK: Cambridge University Press.

2

The Origins of John Harrison's 'Pendulum-Clock' Technology

Andrew King

Within a single ten-year period, 1720–1730, John Harrison (1693–1776) established the fundamental science that not only provided the touchstone for his future life's work but, in a grander scheme, introduced an entirely new technology that resonates with us today.

The Harrison family was literate, talented, skilled and a substantial presence within the local community. They originated from the village of Foulby in the parish of Wragby in the West Riding of Yorkshire not far from Wakefield. Foulby was a part of the estate of Nostell Priory, the seat of the Winn family, who also owned lands and properties in North Lincolnshire. John Harrison's grandfather, Henry Harrison (1632–1701), was parish clerk at Wragby and quite clearly well respected by the Winn family. In 1694, the Nostell estate passed to a new generation. Sir Rowland Winn (1675–1722), aged just 19 at the time of his inheritance, rapidly developed his holdings including the building of a substantial residence, Thornton Hall at Thornton Curtis in North Lincolnshire—a manor held by the Winn family since 1627. This property is on the northern borders of Brocklesby Park, home of the Pelhams, just one-and-a-half miles south of Barrow-upon-Humber. The Winns and the Pelhams likely knew one another, if only through links between the two families through marriage. When a new parish clerk was required in Barrow around 1696–1697, it is perhaps not too surprising that the Winn influence probably resulted in the appointment of John Harrison's father, Henry Harrison (1665–1728), son of the parish clerk at Wragby, thus maintaining a family tradition. The result, of course, was the Harrisons' move to Barrow-upon-Humber (Fig. 2.1) where they would be established for more than a century.

King, A., *The Origins of John Harrison's 'Pendulum-Clock' Technology* In: *Harrison Decoded: Towards a Perfect Pendulum Clock.* Edited by Rory McEvoy and Jonathan Betts, Oxford University Press (2020).
© Oxford University Press. DOI: 10.1093/oso/9780198816812.003.0002

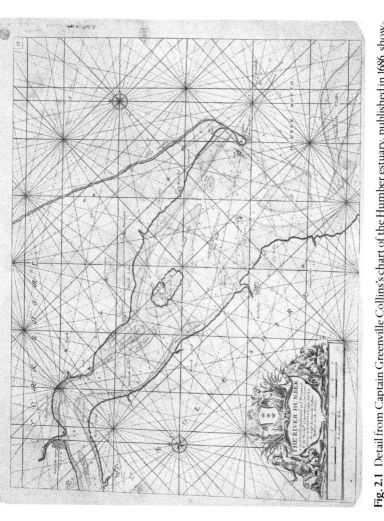

Fig. 2.1 Detail from Captain Greenville Collins's chart of the Humber estuary, published in 1686, showing the village of Barton (around 3 miles from Harrison's North Lincolnshire home). Credit: NMM.

The Harrisons are best known as carpenters and joiners, but a closer look at their lives reveals much more about the range of their abilities than what we know and what we can infer from extant records. Apart from the general work of carpentry around the village, which would have embraced everything from fencing repairs to timber work in the construction of houses, and from making coffins to maintaining water and windmills, the family were also involved in furniture-making and their output included the fitting out of churches. It would appear that John Harrison was also a wheelwright, which included making farm carts and implements such as ploughs and harrows, and which was generally the most respected of the wood craftsmen in village life (Monod-Cassidy, 1980:88).

With all these credentials the Harrisons would have had much esteem within the community. However, John Harrison did not stop there. He also established himself as a land surveyor. The existence of a plane table surveying compass dated 1718, the very year he was first married, is clear evidence of his mathematical ability. Surveying required a fluency in geometry, trigonometry, the use of logarithms and of course basic mathematics (Garnier and Carter, 2015:438–440). All the knowledge Harrison needed for surveying would have been valuable and formative groundwork for the future. If all of that experience wasn't enough, Harrison was fascinated by timekeeping, which became the central theme of his life.

Harrison's interest in clockmaking began during his teenage years; however, during this time he was not a full-time clockmaker. The first clocks he made were mostly of an all-wooden construction. Although the wheels were of oak, the pivots were steel, which ran in brass bushes inserted into the oak plates. These striking clocks could well have been for members of the family or close friends. Harrison's earliest surviving clocks are dated 1713, 1715 and 1717, suggesting that each was made over a two-year period, and it is quite likely that this represents all the clocks produced during this period. By 1718 Harrison was in any case evidently involved in land surveying as well, quite likely working for Sir Rowland Winn and maybe even Sir Charles Pelham of Brocklesby, amongst others (King, 2006b).

Around 1720, however, Harrison's clockmaking skills were brought back into focus when he was commissioned by Sir Charles Pelham to make a clock for the estate at Brocklesby, which today is still the most extensive estate in north Lincolnshire. The hall at Brocklesby was substantially rebuilt during the first years of the eighteenth century, involving a considerable extension of the very substantial stable block, where a clock was required. It is very likely that Harrison gained this prestigious work through the influence of Sir Rowland around 1720 and certainly before early in 1722 when Winn died. This commission was undoubtedly the first of two major turning points in Harrison's life. The second was to occur six years later (King, 2006b:496–512).

At Brocklesby, there is all the evidence of real innovative thought. Following on from Harrison's earlier work, this striking clock (Plate 3) was made largely of oak, including all the wheels except the brass escape wheel. All the solid leaf pinions were of cast brass, made from patterns that Harrison would have produced. All the wheel pivots were of brass running in bushes of the sapwood of lignum vitae (LV), resulting in wheel trains that ran entirely dry, needing no extra lubrication. Lignum vitae (*Guaiacum officinale*) is well known for its extreme hardness; one of the heaviest of hardwoods, it is denser than water. The very dark heartwood contains natural resins which never dry out and provide permanent lubrication. The interesting point here is that while Harrison used LV for the very first time for the pivot bearings, he chose to use the sapwood. It is only the heartwood that contains the natural resins. This can easily be demonstrated by heating samples: the resin clearly rises to the surface in the heartwood, but a similar heating of the sapwood produces no resin at all—it is quite 'dead'. Also, even a simple test proves very readily that the heartwood is noticeably harder.

Harrison had trouble with this clock's lubrication. From the surviving evidence, it is certain that the clock originally had an anchor escapement that was very similar in its layout to a surviving schematic by Harrison (Plate 4). The drawing, although undated, is probably from much later in his life and is likely to be the escapement he made for the clock at Trinity College, Cambridge in 1755. The drawing shows an anchor recoil escapement with an equal lift providing similar impulse from each pallat, and the concave face of the left-hand exit pallat is evidence of the achievement. The problem is that this design leads to considerably more friction on that exit pallat.

A vacant 'mortice' on the right-hand-side of the brass frame (Plate 5), now adjacent to the entry pallat of the grasshopper escapement, follows the same lines as the mortice that is depicted in the drawing. Brocklesby's is the very first grasshopper escapement. The mortice remnant is a fortunate survivor but its existence, while essential to the understanding of the history of the clock, remains because it provides balance (poise) for the pallat frame assembly. The pallats, like the wheel bearings, are made of LV sapwood. The pendulum is a plain brass rod with a very long suspension spring.

Here also is the very first example of Harrison's adjustable suspension cheeks. The brass suspension spring is a full 8-inches long and a ½-inch wide but the acting faces of the cheeks are less than ⅛ inch wide. Harrison emphasised the importance of the cheeks to 'preserve the spring from its ever breaking' quite apart from their contribution to the timekeeping system (Harrison, 1775:45). The brass content of the pendulum rod and suspension spring would have demonstrated to Harrison the temperature effects on the rate of the clock. The temperature of the clock chamber over the stables reaches extremes: in winter, it can be bitterly cold and in high summer it can be insufferably hot. In cold weather, considerable condensation would build up on all the metal surfaces and in summer, the temperature was further increased by the horses below. These extremes would have provided the most profound lesson demonstrating the necessity of providing temperature compensation if any progress was to be made in the accurate timekeeping of clocks. Harrison may have become aware of this from 1723 when the clock would have been running for a short time. However, 1725 was a year of extremes. Agricultural records show that the particularly hot summer was followed by an extremely cold winter (Stratton and Whitlock, 1969:70).

When Harrison installed the new grasshopper escapement he fitted a roller pinion, with live LV sapwood rollers replacing the brass-leafed pinion. None of the other pinions were changed. Another first for Harrison was the use of anti-friction rollers; here just a single roller directly below the front and back pivots. Later Harrison was to use these anti-friction rollers to much greater effect. With Brocklesby, Harrison achieved the design for a mechanical clock to run long term with little or no maintenance and with the grasshopper escapement in its embryonic form—an invention born out of necessity.

The Estate Clock at Brocklesby is uniquely signed 'Harrison Barrow'. It is likely that this suggests a family project: John, the undoubted pioneering spirit, but James (1704–1766)—although only sixteen in 1720—was surely also involved, and quite likely their father, Henry, as well. Whether John's elder brother Henry (1702–1729) had any hand in the construction is not known. This Henry is never mentioned and remains in the deepest shadow of all the brothers, sadly dying at the early age of twenty-seven.

After almost 300 years of running, there is no appreciable wear anywhere in this clock. Brass pivots in LV bushes, whether the softer sapwood or heartwood, remain an ideal bearing combination. The situation is the same with the later pendulum clocks and even with the three sea clocks (H's 1, 2 and 3) that run continuously at the Royal Observatory, Greenwich, UK.

Harrison knew he had struck on something important, something with potential. The Brocklesby Clock is undoubtedly the very cradle for all Harrison's future work on both land-based pendulum clocks and marine timekeepers. In later life, Harrison said that everything he made evolved from what he had built previously. Although the three first clocks were a stepping stone, in reality the Brocklesby Estate Clock is the true foundation.

The experiments at Brocklesby—the learning curve—continued until 1725, at which time Harrison was ready for the next step forward, which was to be truly groundbreaking. In 1725, Harrison produced the first drawings for the next all-important evolutionary phase, and the development from Brocklesby is all too evident. Initially it is tolerably clear that these long case clocks, sometimes referred to by Harrison as his 'Pendulum-Clocks' (a form used in Harrison's (1763) manuscript that will be adhered to here, as is Harrison's spelling of 'pallats'), were a commercial venture. Three of them survive and there is evidence of the start of at least another one and possibly two more, all made to the same design pattern: a batch production of what would be a unique and unprecedented clock design.

The ebonised pine cases were originally similarly decorated with floral artwork in gilt, silver and colour. The dials were a unique combination of softly polished brass chapter rings mounted on oak dial plates, and the plates again ebonised with an all-gilt floral pattern. The overall decorative design of these clocks with their interesting gilt artwork on a dark background was intended to scintillate in the live flame of candle light.

As with the cases, the movements were all principally the same: two train, hour striking with calendar indication and mostly of an all-wood, oak construction. The wheels, apart from the escape wheel, are once again of a composite oak construction. The wheel teeth grouped in segments let into the deeply slotted periphery of the wheels—clearly inspired from windmill gearing.

Harrison would have known that these wheels would shrink across the grain, leaving them out of round. There were now roller pinions throughout. The wheel trains were designed so that any ovality of the wooden wheels would not affect the mechanical running of the wheel trains. The pinions are all like the escape pinion at Brocklesby: brass bobbins with live (free-moving) LV rollers on brass pins.

This time Harrison used the heartwood of LV throughout. The wheels with their mating pinions are set on chordal pitch centres rather than the normal circular pitch centres. The extra deep engagement only permits the pinion rollers to roll on the flanks of the wheel teeth, never engaging the wheel tips. With the development of LV anti-friction wheels grouped in pairs for both escape wheel pivots to accept both load and thrust, the overall running of the wheel trains of these clocks has much-reduced friction and all contact is rolling, not sliding (except on the fixed brass pins of the a/f wheels and the pinion rollers). The freedom of the pinion rollers is essential for the grasshopper escapement. A solid pinion would experience an unacceptable level of friction owing to the escapement's recoiling action, advancing and slightly reversing the wheel train at every beat of the pendulum.

Initially the layout of the escapement of these clocks was probably as at Brocklesby (Plate 6), with the two independently pivoted pallats acting in compression (the 'double thrust' version of the escapement). Whilst the original escapements of the first two clocks are lost, in Harrison's clock No. 2 the remnant of the original grasshopper pallat frame remains. Most fortunately when this clock was converted to an anchor escapement, possibly in the nineteenth century, the new anchor pallat was sculpted to fit the remnant of the grasshopper frame. This relic is possible evidence of the original existence of the Brocklesby variety. In clock No. 3 (in the collection of the Worshipful Company of Clockmakers, London), there is evidence that this too was originally fitted with the early double-thrust version. When clock No. 2 was under conservation and restoration, just such an original type of escapement was returned to the clock, in interpretation of this evidence.

Sometime later, possibly after Harrison moved to London, he revised the escapement, bringing both pallats and composers together, with all four moving parts now on a common axis. This is John Harrison's co-axial grasshopper escapement (Plate 7). It is an impressive example of his innovative and progressive thought.

Finally, but crucially important to the integral design, is the pendulum system. It would appear that it took Harrison at least another two years or so from the start of the project in 1725 before he finally solved the problem that resulted in the gridiron pendulum (Short, 1752).

Writing in 1730, Harrison remarked: 'But there is no wire of any metal, whatever, whereof to make a pendulum but what is continually altering its length according to the degrees of heat and cold; and this I discovered above 2 years ago'. Taking Harrison at his word this would mean that the design was complete before June 1728. To conclude the experiments, and there must have been many of them, Harrison needed cold weather and so potentially, this would have taken place during the previous winter of 1726–7 or maybe the months of May and June of 1728. In these early summer months the mornings and evenings can be quite cool in contrast to the warmth of the days. Also (still in 1730), Harrison talks of the clock he sold 3 years ago. Referring to the middle of 1727, he continued to say that 'it did not have a pendulum as truth requires'—a suggestion that at this date the pendulum was still in an embryonic state (Harrison, 1730).

It would now appear that, having passed through several prototype stages, the fully developed gridiron pendulum (Plate 8) probably evolved just before 1728. This remarkable and entirely original invention was achieved by research and though painstaking observation and trial. Harrison's gridiron pendulum is a very slender thing, just 1⅜ inches wide. The wires are 0.080 inches for the steel and 0.090 inches for the brass, the thicker brass employed to ensure an equal rate of expansion, owing to its greater conductivity of heat.

Experiments were carried out with varying thicknesses of metals as well as brass from three different sources: London, Sheffield and Holland. The tin whistle adjuster for the fine adjustment of the ratio of the brass to steel wires was fitted in a break of the final steel central wire, close to the lower end of the gridiron assembly receiving maximum air flow and thus sensitivity to the environment. The first two of these clocks were completed by the end of 1726, displaying an astonishing performance.

Harrison described the trial methods in 1730 in some considerable detail:

> After I discovered ye wire to be longer and shorter by Heat and Cold, I prepared a convenience on the outside of my house where ye sun at 1 or 2 o'clock makes it very warm to try different quantity ye one foot of metal altered in proportion to another. (Harrison, 1730)

He follows this with a table of the expansion rates of the metals involved. To have moved from the realisation that metals expand and contract with temperature changes to discover that different metals change by different amounts, and to be able to establish the ratios, is truly remarkable. Perhaps Harrison fitted one of the clocks in the convenience itself (believed to be a glazed porch adjoining his house that experienced extremes of temperature and facilitated this testing) to begin with, taking maximum possible advantage of the early summer days.

The development of the gridiron pendulum is one of Harrison's greatest achievements, but this highly motivated pioneer always seemed to know where he was heading, taking the enormous amount of work involved in his stride. The process of adjusting the clocks to bring them to keep time was no mean feat. Harrison achieved a rate of accuracy and stability that was quite astonishing for the eighteenth century. Although his claims were treated with the greatest scepticism over the years that followed, overall Harrison's testing method was unique, systematic, and comprehensive.

Harrison's explanation of how he derived a time standard is lucid and graphic.

> I have two ways of trying the Motion, ye which together is very compleat. One is by the apparent motion of ye fixed stars with a very large sort of an instrument of about 25 yards radius composed of the East side of my neighbours chimney (which is situated from my house towards ye South) and ye west side of an exact place of some one of ye upright parts of my own Window Frames; by which ye rays of a Star are taken from my sight almost in an instant. And I have another person to count the seconds of ye Clock, beginning a little before ye Star vanish. So I observe what second is mentioned when it vanishith; & I have a table calculated to show how much sooner any such Star is to vanish every night before the 24 hours of ye pendulum day is expired. (Harrison, 1730)

Harrison had created his own transit instrument. This method was well established and was known to both astronomical circles and the clockmaking fraternity. For instance, it is described by John Smith in his 1694 publication, *Horological Disquisitions*, and it is also shown by Jeremy Thacker in his Longitude pamphlet of 1714.

The second part of the 'ye completion' is the adjustment of two clocks against his established time standard. This method was far in advance of anything else of the time and Harrison eloquently described this intriguing procedure:

ye two clocks placed one in one room and ye other in another, yet so, that I can stand in ye doorstead and hear the beats of both ye pendulums, when ye Clock Case Head are off; & before or after the hearing can see ye seconds of one Clock whilst another person count ye seconds of ye other: by which means I can have ye difference of ye Clocks to a small part of a second. And in very Cold & frosty weather, I sometimes make one room very warm with a great fire whilst ye other very Cold. And again the contrary. And sometimes ye like in Summer by ye Sun's Rays in at the Windows of one room and also a fire whilst ye other is close shut up and Cool. Thus I prove ye operation of ye Pendulum Wires & adjust ye same...And to prove or adjust ye Cycloid to Vibrations performed in different Arches as required...I cause ye Pendulum to describe such by increasing & decreasing ye draught of the Wheels and that by adding or taking from the Weight by which I can make 8 or 10 times more difference than Nature ever will and yet the effect be nearly the same...as if Nature itself had altered the weight of the Air so much. (Harrison, 1730)

Both clocks were equally adjusted using one as a control to make alterations to the second clock, the subject, and then—which is almost lost in the text—'and again the contrary', the roles are reversed so that the second clock becomes the control. Harrison described the location of the two clocks on test:

Mine are in low rooms which are often without fire a fortnight or 3 weeks together in very bad weather & yet I can discern no alteration in them, I sold one 3 years ago which stands in an upper room where there is seldom any fire & its pendulum describes the same arch now that it did half a year before I sold it [this referring to late 1726] and so I believe will continue to do a long time. It is made...with an imperfect cycloid but not such a pendulum as truth requires. (Harrison, 1730)

This is a reference to clock No. 2 that was sold in 1727. Evidence from clock No. 2 indicates that the cycloid cheek plate has indeed been removed. Perhaps, therefore, when Harrison mentioned an 'imperfect cycloid' it was a reference to his realisation that the suspension cheeks for his clocks needed to be of circular form and not cycloidal, and that No. 2 originally had an interim cheek form when he thought that the cheeks still needed to be cycloidal. The present existing pendulum suspension plate on No. 2 has a simple cock with no cheeks at all, so on close examination, it is reasonable to suppose that rather than leave the 'imperfect cycloid', this replacement may well have been made by either John or James. Whilst these clocks appear to have been made for commercial reasons, something happened that caused a dramatic change of direction in Harrison's clockmaking.

The second and most profound turning point in Harrison's life was when he heard of the 1714 Longitude Act, which the clockmaker indicated he first heard of in 1726 (Harrison, 1770). According to the Act, to secure the ultimate award of £20,000 the successful method had to prove to be accurate to within half a degree of longitude after a voyage to the West Indies—an accuracy of within three seconds a day or better. To be of use, the method would have to have a consistent stability over long periods as ships could be at sea for many months and often years. Once aware of the Act, Harrison made it very clear that from that point onwards all other clock and watch activity would be given up to pursue the longitude goal. However, as a reliable workshop regulator was a prerequisite, work on the development of his pendulum clock concept continued. The earlier work on pendulum clocks was the basis for the evolution of the forthcoming sea clocks.

From his earliest surviving writing, a document specifically dated 10 June 1730, Harrison carefully describes his pendulum clocks, their construction and his testing methods. Harrison's style of writing deteriorated substantially in his twilight years. *A Description Concerning Such Mechanism*... does not share the same lucidity of the 1730 manuscript; it is complicated, disjointed, and verbose, and lacks punctuation. His final published work from 1775, although full of crucial technicalities, is badly written and poorly structured. It was heavily criticised in the press (Griffiths, 1775) and also by his grandson, John Harrison (1761–1842), who, writing under the anagram 'Johan Horrins' in 1835, observed:

> that modesty for which John Harrison was conspicuous in his bright days had forsaken him when in 1775 at the age of 82 he published his pamphlet, "A Description of Such Mechanism ...". He would accept of no assistance in revising this work which is so encumbered by his singular and undefinable manner of expressing himself in writing as to be unintelligible to the general reader without translation and although it has valuable hints for those who are thorough masters of the subject it may be set down altogether as a Memento Mori. (Horrins, 1835)

There is here, at least, a concession to 'valuable hints'. This is true and important, even if the volume is dismissed pejoratively as a mere pamphlet. Only the most persistent reader will see that Harrison was an original, independent, free-thinking pioneer. Although stubborn, he pursued his ideas with an obsessive driving passion, and at the same time, through the light of pure reason, he could ruthlessly abandon work proven to be flawed and move forward with a new understanding—the hallmark of an engaging scientist on a determined quest.

Harrison's pendulum technology, offering the possibility for a far greater and stable timekeeping accuracy than any clock made by his contemporaries, effectively died with him in 1776. It is only now that Harrison's pendulum practice has finally been put to the test and we can begin to appreciate what Harrison was trying to tell us around 240 years ago. The world of technology always pushes the boundaries: to climb higher, to travel further and faster—and the history of accurate timekeeping is no exception.

That earliest surviving document of 1730, presented in 29 sections, was concluded with an unequivocal statement of authorship, dated with Harrison's full address. There is no surviving title page, but it seems likely that it could have been the draft for a longitude proposal that did not, in the end, need to be published. By 1730 Harrison almost certainly had the beginnings of the support he needed, starting with his mentors, the Astronomer Royal Edmund Halley (1656–1742) and George Graham (c.1673–1751). The document begins by outlining the experiments with pendulum clocks and proceeds to the proposal for adapting the design for use at sea, a timekeeper that was probably already under construction:

> Some years ago I made several alterations in order to render ye motion of Clocks more exact than heretofore, but when I came to try them by strict observation...I judged ye best performance of ye best Pendulum Clock I ever saw, made or heard of to be incapable of this matter. Would it go as well in a ship at sea in any part of ye World as in any one fixed place on ye land. Yet from several observations, I still endeavoured to make further corrections in this Motion; and in these 3 last years have brought a Clock nearer ye truth than can be well imagined, considering ye vast Number of seconds of Time there is in a month, in which space of Time it does not vary above one second, & that mostly ye way I expect: so I am sure I can bring it to ye nicety of 2 or 3 seconds in a Year. And 'twill also continue this exactness for 40 or 50 years or more; however so as not to vary above 2 or 3 seconds from what it did ye Year next before; for 'twill not want cleaning, & the little it wears can alter it but insensibly little. This nicety is owing partly to ye matter ye Clock is made of, partly to ye contrivance it is made with, & partly to ye nice observations it is try'd by, & ye convenient place it stands in. (Harrison, 1730)

The most remarkable point here is the claim that he achieved a rate of accuracy 'not varying above a second in a month' and, furthermore, that the Clock would run for '40 or 50 years or more' without cleaning. Writing 30 years later, Harrison claimed:

> (after the adjusting is completed and as in supposing the Clock not to be removed) for a 100 or 200 Years without any cleaning (viz as upon my materials and construction) and that to a nicety of a ¼ of a second or less in a month or perhaps as taking any Year throughout to less than 2 seconds supposing without the Clock to stand in a pretty temperate place and the case to be very good and close as that no dust or cobwebs may get or increase on the pendulum. (Harrison, 1763)

Once again, Harrison emphasises with considerable confidence the breadth of his achievement and his ultimate goal of accuracy to within 'less than 2 seconds' in a year—a self-assurance that he was to maintain unwaveringly for the rest of his life. By the time Harrison became aware of the Longitude Act he already knew that he had the basis of a timekeeping system that was far superior to anything known and which could certainly meet the exacting demands of the Act. The challenge was to incorporate this technology into a portable timekeeper.

Harrison's claim of accuracy for the 1720s wooden Pendulum-Clocks, whilst generally doubted, has never been put to the test. Everything about these clocks contradicts all that is held sacrosanct in the world of traditional clockmaking—a tradition that arguably stretches from Christiaan Huygens's published explanation of the pendulum clock through to the variations and developments of clocks by Shortt and Fedchencko in the twentieth century. But Harrison remained adamant about the performance of his clocks and their running to within 'a second in a month', which was even reiterated by Martin Folkes in 1749 in front of an audience at the Royal Society when awarding Harrison with the Copley Medal—the highest accolade of the society.

The remarkable performance of Clock B, made by the horologist and sculptural clockmaker Martin Burgess, and which has been on test at the Royal Observatory, Greenwich since 2012, is due to John Harrison's advanced thinking up to his death in 1776. The system evolved from the embryonic Pendulum-Clocks made in Barrow some 50 years earlier. A programme is now under way to make a pair of the early Pendulum-Clocks to test them as described by Harrison and to prove that his remarkable rate of accuracy is indeed achievable.

It is important to summarise the key elements of Harrison's early clocks.

- The wheel trains are entirely new to horology. The main wheels made of oak are designed so that any shrinkage will have minimal effect on the mechanical running. The pivots run in bushes of LV and require no extra lubrication. The movements (if kept dust-tight) are designed to run for hundreds of years with no maintenance.
- The grasshopper escapement is a new invention and does not require lubrication.
- The gridiron pendulum and the suspension system is a new invention.
- The testing methods together are a new and radical approach contributing to the remarkable claimed rate of accuracy.
- In 1725 John Harrison redesigned the Pendulum-Clock—an entirely new concept.

The pendulum technology that could have produced an unprecedented standard of timekeeping in the later eighteenth century was not reached until the twentieth century.

Harrison revolutionised the design of the pendulum clock, invented by Christiaan Huygens about 1656. Had it been embraced by his contemporaries, it may well have substantially accelerated the advancement of science and horology. All, however, was forgotten and on the 24 March 1776, this revolutionary system effectively died with him.

References

Garnier, R. and Carter, J. (2015). *The Golden Age of English Horology: Masterpieces from the Tom Scott Collection.* Winchester, The Square Press.

Griffiths, R. (1775). *The Monthly Review or, Literary Journal,* vol 53, October 1775, 320–329.

Harrison, J. (1730). *The 1730 manuscript.* Refer to Bromley, J. (1977) *The Clockmakers' Library: the catalogue of the books and manuscripts in the library of the Worshipful Company of Clockmakers.* London, Sotheby Parke Bernet Publications, P.108. MS977.

Harrison, J. (1763). *An explanation of my watch.* Available at https://ahsoc.contentfiles.net/media/assets/file/Explaining_My_Watch.pdf. Accessed 4 December 2018.

Harrison, J. (1770). *The case of Mr John Harrison.* London.

Harrison, J. (1775). *A Description Concerning such Mechanism*. London, Jones, T. https://ahsoc.contentfiles.net/media/assets/file/Concerning_Such_Mechanism.pdf. Accessed 12 December 2018.

Horrins, J. (1835). *Memoirs of a Trait in the Character of George III*. London, W. Edwards.

King, A. (2006a). John Harrison Clockmaker, Part 1: His first longcase clocks 1713–7. *Antiquarian Horology*, 29(4), 475–488.

King, A. (2006b). John Harrison Clockmaker, Part 2: The Brocklesby Turret Clock, c. 1722. *Antiquarian Horology*, 29(4), 496–512.

Monod-Cassidy, H., ed. (1980). *Journal D'un Voyage en Angleterre 1763*. Studies on Voltaire and the Eighteenth Century. Oxford, The Voltaire Foundation.

Short, J. (1752). A letter of Mr James Short, FRS to the Royal Society, concerning the Inventor of the contrivance in the pendulum of a clock, to prevent the irregularities of its motion by heat and cold. *Philosophical Transactions of the Royal Society*. London, Royal Society, 517–524.

Stratton, J. M. and Whitlock, R. (1969). *Agricultural Records AD.220–1968*. London, J. Baker.

3

Introducing Martin Burgess, Clockmaker

William Andrewes

If you had told someone during the 1960s that in fifty years mechanical wristwatches would be in high demand; that the story of finding longitude at sea would be the subject of a best-selling book, a television drama, a documentary, and a touring exhibition; and that a mechanical clock with a pendulum swinging in free air designed on principles proposed 200 years ago would keep time to a second in 100 days, you would have been regarded as a lunatic and politely ignored. For at that time, the clock- and watchmaking industry had been in a steady decline for years, and, with the advent of quartz, the only future for aspiring students of horology appeared to be in restoration or the antiques trade.

Among the very few who envisaged a brighter future—a renaissance of mechanical timekeeping—were the watchmaker George Daniels and the clockmaker Martin Burgess. While Martin greatly admired what George was doing for the future of watchmaking, George could not comprehend what on earth Martin was doing trying to prove that John Harrison's claims of accuracy were true. Harrison had written in his 1730 manuscript that his wooden regulators kept time to a second a month, two or three seconds in the course of a year, and this accuracy was clearly stated in 1749 when the President of the Royal Society awarded Harrison their highest honour, the Copley Medal. Just like Harrison's later (1775) claim of a second in 100 days, this extraordinary achievement was dismissed as a complete exaggeration. George Daniels made no bones about saying that the very idea of such accuracy for a wooden clock was absurd and that a second in 100 days for any kind of mechanical clock should be ignored as a mistake. Even Colonel Humphrey Quill, the leading historian on the life and work of John

Andrewes, W., *Introducing Martin Burgess, Clockmaker* In: *Harrison Decoded: Towards a Perfect Pendulum Clock.* Edited by Rory McEvoy and Jonathan Betts, Oxford University Press (2020). © Oxford University Press. DOI: 10.1093/oso/9780198816812.003.0003

Harrison during the 1950s and 1960s and author of the 1966 book, *John Harrison: The Man Who Found Longitude*, who introduced me to George, wondered if perhaps there might be some exaggeration in Harrison's writings. There was nothing to prove that Harrison's claims were true. Only the word of an honest man, whose genius was founded in the search for truth: a man who was known to start again when a theory proved false in practice. Neither Daniels nor Quill lived to witness the potential for Harrison's designs, verified beyond a shadow of doubt by a clock designed and made by Martin Burgess on the principles Harrison had prescribed: in a carefully supervised trial at the Royal Observatory, Greenwich, starting on 20 March 2014 and ending on 7 July 2016, the error of Burgess Clock B from Coordinated Universal Time was found to be one second fast on account of a leap second having been intro-duced during the trial period. When it was finally stopped in the presence of the same distinguished committee that had sealed its case at the outset of the trial, the most it had ever drifted during this period of 840 days was a fraction over two seconds slow in the heat of the summer of 2014. It is extraordinary that no one had ever made any thorough attempt to understand or to duplicate Harrison's regulator science before.

Like Harrison, who devoted most of his life to the perfection of preci-sion timekeeping both on land and at sea, Martin Burgess has spent much of his trying to understand Harrison's technology and prove its merit to an unbelieving world. He knew from experience that the only way he could substantiate Harrison's claim both to himself and to the outside world was to build a clock based on the general principles described in *A Description Concerning such Mechanism* (Harrison, 1775). That Martin—also like Harrison—had no formal training as a clockmaker may have been an advantage, allowing him to approach clockmaking from his own unique perspective. This chapter provides insight into Martin's life and work before the formation of the Harrison Research Group, a team of dedicated horologists and friends devoted to under-standing Harrison's writings and his regulator science. This group, founded at the suggestion of Mervyn Hobden, was crucial to the design and construction of Martin's two regulators: Clock A (commissioned by Gurney's Bank, Norwich) and Clock B, the subject of this volume that underwent the above-mentioned trial at the Royal Observatory, Greenwich. Martin would never want to give himself credit for any-thing that has happened, preferring to say that it was the research group as a whole that made it all possible. But without Martin's dogged perse-verance and enduring faith that he was on the right path, there would be no practical proof of the brilliance of John Harrison's regulator sci-ence. For this reason, in July 2014, along with the events celebrating the tercentenary of the Longitude Act, Martin was honoured by being chosen as the first recipient of the Derek Pratt Prize, offered by the Worshipful Company of Clockmakers in memory of the accomplished English horologist Derek Pratt (1938–2009)—one of the most distinguished and valuable prizes offered in the craft and science of time and timekeeping.

Martin Burgess, named 'Edward Burgess' on his birth certificate, was born on 21 November 1931 in Sculcoates, a town in Yorkshire about two miles north of Hull. (Hull is situated right across the Humber Estuary from Barrow-on-Humber, where John Harrison lived and made his first clocks.) His father came from a wealthy banking family, and his mother was the daughter of a bricklayer. They were married one month before Martin was born, but, because his father's family considered their new daughter-in-law such an inappropriate match for their son, their baby was sent to an orphanage. By good fortune, when he was about six months old, 'Edward' was adopted by Cicely Pelly, a spinster from a well-to-do family who lived in a large country house called Witham Lodge, about nine miles northeast of Chelmsford in Essex. Cicely's father, Edmund Pelly, would never have allowed this, but his death seven weeks before Martin was born left his wife and his unmarried 40-year-old daughter living alone. Cicely did not like the name Edward, so she added Martin as a middle name, and this name was then used.

Both of Cicely's parents had inherited money from their families. The majority of their wealth came from Cicely's maternal grandfather, John Fowler of Leeds—a well-known agricultural engineer who pion-eered the manufacture and use of steam engines in ploughing and digging drainage ditches. The Pelly family's secure financial situation enabled them to run Witham Lodge—unlike many country houses that were struggling to survive at that time—in a traditional manner, with three housemaids, a cook, a parlour maid (who served at table), a gardener, a chauffeur, and, after the adoption, a full-time nanny. Thus, Martin enjoyed a privileged upbringing and was assured of an excellent education. His adoption by Cicely Pelly was the first of many acts of the 'unseen hand' that has blessed Martin throughout his life.

Being left-wing and very independently minded, Cicely Pelly did not like Witham Lodge or its outdated lifestyle, so in 1936 she purchased several acres of land about six miles southwest of Witham Lodge on the outskirts of the village of Boreham. Here, she built a new house overlooking farmland that sloped gently down to the meandering River Chelmer. Mulberries, as Cicely chose to call her and her son's new home, was completed in 1938, the year before Cicely's mother died (Plate 9).

Martin was sent to a local kindergarten and then, when he was eight, to King Edward the Sixth Grammar School in Chelmsford. Leaving his bicycle at the local wheelwright's forge before catching the bus to school gave him his first exposure to blacksmithing, a skill that would later become an integral part of his sculptural approach to clockmaking. In 1944, he was sent to Gresham's, a boarding school in Holt, Norfolk. With its emphasis on the arts and sciences and learning by doing and making, Gresham's played a significant role in shaping his future. While there, Martin made his first clock, filing all the wheel teeth by hand since no wheel-cutting engine was available. Years later, he would go on to remake the movement when he took up a place at the Central School of Arts and Crafts. He also developed a keen interest in chain mail, especially in how it had originally been made. Unable to satisfy his curiosity on this subject, one of his teachers suggested that he should write to Sir James Mann, the Master of Armouries at the Tower of London and a leading expert on medieval and Renaissance armour. Martin's letter was well received, and the meeting duly took place in London. Sir James was astonished by the genuine interest and insights of this 16-year-old boy, particularly when Martin pointed out that the size of the links varied at the joints to permit greater freedom of movement. Apparently, no one had done an in-depth analysis before on how chain mail was made, so, encouraged by Mann, Martin became one of the first to study its construction in detail. As a result of this research, Martin was later invited to join the Society of Antiquaries. During his time at Gresham's, Martin also made a copy of a sixteenth-century bellows-visor bascinet, which he formed out of copper sheet. On applying to art school when he left Gresham's, it was his bascinet in particular that impressed the faculty at the Central School of Arts and Crafts (now called the Central School of Art and Design), in Holborn, London, and secured him a place in their three-year silversmithing course.

Soon after starting at Central in 1950, Martin came into contact with two great craftsmen who had an enduring influence on his appreciation of design and craftsmanship: Francis Adam, a blacksmith who could transform an old sheet of iron into a beautifully curled oak leaf, and Bert

Brooker, a wonderful mentor and superb silversmith who taught Martin to become an expert in silver soldering and in using a piercing saw—techniques that he later employed extensively in making sculptural clocks. One of the things that Francis Adam pointed out to Martin was that few straight lines exist in nature. Even though some lines may look straight, they are not—and that is part of their intrinsic beauty. That both Francis and Bert saw and practiced their craft as an art helped Martin to see clockmaking in the same light: from this time on, Martin would dispense with the clock case and design the movement—the frame, the driving weight, the wheels and pinions, the escapement, and the pendulum—as elements of sculpture in their own right.

In 1953, shortly before Martin completed his course at Central, Professor Anthony J. (Tony) Arkell, the curator of the Flinders Petrie Collection at University College, London, visited the school in search of a good student to assist him as a technical assistant with the conservation and restoration of Egyptian antiquities. Delighted by the prospect of such a unique opportunity, Martin applied for the job, and, after graduation, went to work for Professor Arkell. His annual salary of £350, which had increased to £800 by the time he left ten years later, was just enough to live on in central London. He learned much at University College from the remarkable Egyptian treasures he helped to care for, in particular the subtleties of the curves and virtual absence of straight lines in Egyptian art. Martin had not been there very long before he was introduced to Tony Arkell's daughter: Eleanor was a very bright, well-educated, independent woman of extraordinary determination, not dissimilar to Cicely Pelly. She was three years younger than Martin, but they had much in common and became close friends. When Tony Arkell retired in 1963 and came to thank and bid his student farewell, Martin responded that it would not be good-bye: he and Eleanor had decided to get married.

Martin never met his birth parents and knew very little about them. Therefore, it came as a complete surprise when, in 1959, he received a letter from the executors of his father's estate informing him that, since his father had died and he was the only surviving heir, he was the beneficiary of his father's inheritance. This substantial windfall, combined with the financial security of his adoptive mother, meant that Martin would never again have to work for a living. He could now devote all his time and energy to sculptural horology.

Despite this financial security, in 1963 when Martin and Eleanor left London and settled with Cicely at Mulberries, they chose to live in a very frugal manner. Around 1950, thinking ahead to the day when she hoped Martin might come back to live at Mulberries, Cicely added a self-contained apartment with its own entrance and kitchen and bathroom on the west side of the house. In contrast to the £600 cost of the original house in 1938, this addition, twelve years later, cost £1,000. Like Martin, Eleanor had a great love for the country, much preferring the disposition of animals to that of most human beings. In addition to dogs, cats, and chickens, the Burgesses acquired several goats, which provided the milk they drank and the cheese and whey they ate. The goat manure was wheel-barrowed from the goats' enclosure to an area adjacent to the vegetable garden and shovelled on the dung heap, which had been carefully designed in a square to the exact specifications of the dung heaps used in ancient Egypt. This produced excellent fertiliser for the vegetable garden, from whence came most of their food. Beekeeping also gave them a plentiful supply of honey. During the summer and autumn, a section of the hedge (a type of plum with memorably sharp thorns) that bordered the property was cut every evening to feed the goats. In winter and early spring, the goats were fed with hay that had been harvested the previous year from the abundant supply of long grass growing in the surrounding fields and stored in haystacks. Among the few things that Martin and Eleanor ever purchased were flour, from which Martin baked their daily bread, and malted barley and hops, from which he made a delicious but deceptively strong beer.

Working the land, feeding the goats, and producing nearly all of their own food were duties that had to be attended to every day, regardless of the weather or personal ailments. There were no holidays from this manual labour, but Martin and Eleanor would not have it any other way. Anyone who came to stay with them was expected to help. The day began at sunrise and ended early, to allow time to cut the hedge or the grass, or attend to anything else that had to be done before sunset. Because of the limited hours of daylight, Martin never had a phone in his workshop. The work he was doing required intense concentration, and, after such a distraction, it took him several hours to get back to where he was and resume his train of thought. With Eleanor out in the fields and no answering machine, the only time to call was after sunset—and then you would be assured a long conversation.

Martin began making his first sculptural clock around 1960, when he was 29. After leaving Gresham's, he had stayed in touch with his old teachers and went back to see them each year, sometimes taking something with him to show them. On one occasion he brought with him a very unusual teapot, the last piece of silver that he had designed and made at Central. Another older Gresham's alumnus, a design critic named Richard Jones who happened to be visiting at the same time, was intrigued by it and asked Martin if he could feature it in one of his articles. Martin was duly invited to Jones's very modern house in Broxbourne, a small town in Hertfordshire about six miles southeast of Hertford. Martin noticed that there was no clock in the house and asked Jones what sort of clock he would put in a house like this. Their conversation led to a number of ideas and a location in the middle of the house. No commission was given, and none expected. Having received a substantial inheritance the year before, Martin did not need the money and was prepared to do it in his spare time at his own expense. He wanted to make a horological sculpture, as he once described it, 'where the mechanics is the design'.

To build the Broxbourne clock, Martin converted an old chicken shed at Mulberries to serve as his workshop (Plate 9). Despite the fact that this was not the perfect space and he still did not have all the tools he needed, he firmly believed that this should not be an impediment. Making the clock took every evening and weekend for three years, working under the most challenging conditions without any machines, so everything had to be done by hand. But the clock had been designed with these limitations in mind. For example, to overcome the problem of having to make properly shaped wheel teeth, he employed roller bearings in the pinions, and to minimise friction, he used ball races at the pivots. The Broxbourne clock was installed in Richard Jones's house in late 1963 (Fig. 3.1). It ran for one year on a single winding with a seven-foot fall for the weight—but took its owner 26 minutes to wind! When Richard Jones sold his house, the Broxbourne clock was returned to Mulberries. It remained there until 1984, when Martin agreed to sell it to Seth Atwood for The Time Museum, in Rockford, Illinois. It was displayed for many years at the entrance to the Clock Tower Inn, the hotel where The Time Museum was located, until it was sold at Tom Harris Auctions in Marshalltown, Iowa, in May 2006. Its current whereabouts is uncertain.

Soon after installing the Broxbourne clock at Richard Jones's house and moving back from London to live permanently at Mulberries, Martin hired an architect to design and build a proper workshop adjacent to the house. He planned this to be fully equipped with all the tools

Fig. 3.1 The "Broxbourne" clock, Martin Burgess's first sculptural clock, shown in the house in Broxbourne, Hertfordshire – the original location for which it was designed and made between 1960 and 1963. Photograph © Martin Burgess.

he needed: on the ground floor, he included a forge, a large anvil, a Myford lathe, a Fobco floor-standing drill, a grindstone, a gas torch with four choices of flame, and workbenches; on a balcony above the forge was a studio with his drawing board and library. A large sliding glass door in the workshop opened onto a path leading to the back garden.

While in London, Martin had made many friends involved in the creative arts. One of them, a record producer named Simon Napier-Bell, asked him in a joking manner to make a sculptural clock that would be so intriguing to watch that, carefully positioned alongside his desk, it would distract his clients when they were reviewing and signing contracts. From this commission came Martin's second sculptural clock, nicknamed 'Simon's Balls' after the large polished brass spheres on either end of an 80-pound compound pendulum that rocked in a mesmerizingly slow two-and-a-half-second period (Plate 12). Part of the magic of this clock is its large grasshopper escapement, a sign that, certainly by 1965 when this clock was made and possibly earlier, Martin was already strongly influenced by the work of John Harrison. In the 1969 film *Clockmaker* by Richard Gayer, Martin mentioned that he must have already been influenced by Harrison's work when he was making the Broxbourne clock, because, as he put it, it was 'chock full of rollers'. The Simon's Balls clock has had a chequered history and may have ended up on the scrap heap had it not been acquired in 1997 by Donald Saff, who engaged Richard Ketchen to restore it to its present fine condition. A half-scale copy of this clock was made by Richard Good as a prototype for a limited production run of thirty-five marketed by E. J. Dent as a limited edition model named 'Concorde'.

In 1966, not long after completing Napier-Bell's commission (the first clock to be made in his new workshop), Martin embarked on a sculptural clock that he envisaged could be made in a limited edition of twenty. Wanting the theme of this clock to be 'time', he collaborated with an artist named Susan Butler, one of Eleanor's closest family friends. 'Peggy', as they called her, designed the clock's three 18-inch-diameter wheels that depict the hare and the tortoise (Fig. 3.2). Martin

Fig. 3.2 Martin, ca. 1971, in his workshop in front of the prototype of the "hares and tortoises" clock, which was displayed in his workshop for many years. Photograph © William Andrewes

then cut the teeth and pierced the design out of duralumin—an age-hardenable alloy composed of aluminium, copper, manganese, and magnesium. Martin used this material extensively in his clocks of this period, because its matte silver-like finish was durable and worked very well visually in contrast with the black frame. The frames of these clocks were made of wrought mild-steel, forged by Martin with occasional help from Eleanor, who could wield a 14-pound hammer accurately all day without tiring. (Martin, who from his own work at the forge was very strong, once said to me: 'If anyone tries to break into our house, it's not me they should be worried about!') With so many parts to make, the work on these clocks was exceedingly arduous and slow, so he ultimately decided to focus on finishing the prototype and to use it to promote the sale of the twenty clocks. Evidence of the intended number of clocks in this limited production run is contained in a letter (undated, but about 1981) that Martin wrote to me when I was trying to acquire one of his clocks for The Time Museum in Rockford, Illinois. Martin told me that the prototype, which was still hanging on the wall of his workshop, was not for sale, but that Stephen Higginbottom was planning to complete the run of twenty and to give Martin a percentage of each one he sold. At the end of the last paragraph, he adds: 'The 20 will have the prototype bugs removed.' However, although the large wheels and the frames were constructed by Burgess, only the prototype was completed: the others remained as parts stacked up in his workshop (where they remain at this time of writing), because the opportunity of a lifetime suddenly emerged.

In addition to many other activities in the local community, Cicely Pelly was a Justice of the Peace on the Witham Bench, and one of her close friends, Esther Evans, served in the same capacity on the Chelmsford Bench. Mrs Evans was familiar with Martin's work, because, in late 1963 when Martin and Eleanor were married and came back from London to live at Mulberries, her son had been chosen as the architect for Martin's new workshop. In 1966, Mrs Evans was at a dinner in London, seated by chance next to the Managing Director of the merchant bank J. Henry Schroder Wagg & Co., whose new premises at 120 Cheapside (not far from St. Paul's Cathedral) were under construction. He told her about the bank's recent move and of his wish to make the lobby of the new building look more attractive. 'Why,' said Mrs Evans, 'you need Martin Burgess to design you a sculptural clock!' Her dinner companion had no idea what that was, but she made it sound so enticing that Martin was duly summoned to London to discuss the possibility.

The guidelines given to Martin were that the clock had to be visible from the street and indicate the time in all the cities where the Schroder Group operated. Seeing this as an excellent opportunity to put sculptural horology on the map, Martin envisaged a clock with a 10-foot-diameter great wheel—the largest of its kind in the world—that rotated once in 24 hours. After finalising the details of the commission, he devoted the next two and a half years to refining the design and constructing all the parts. It was a massive undertaking to build in his small workshop, but he refused to let anything stand in his way. He made every part himself (Plate 10), with Eleanor's help from time to time. The great wheel has 1,440 teeth, one for each minute of the day, and weighs 70 pounds; the pendulum weighs 170 pounds and swings once every one and a quarter seconds (Plate 13). Its installation was no easy task either, and for this, he enlisted the help of several old friends, including his old master Bert Brooker.

The Schroder clock, which was completed in 1969 just before Martin's 38th birthday, attracted a good deal of publicity, including Richard Gayer's documentary film *Clockmaker*. The film gives a wonderful portrayal of Martin's life at that time: not only the extraordinary achievement of making and installing the Schroder clock, but also a fascinating insight into Martin and Eleanor's life at Mulberries. Martin's story also appealed to a writer from *The Guardian* whose article 'Clocks and Goats' Milk' was published in January 1970 (Stevens, 1970). These were my first introductions to Martin's life and work. Intrigued by what he was doing and delighted to find someone who shared my interest in John Harrison (as nascent as it was at that time), I plucked up courage and wrote to him. He responded almost immediately, inviting me to Mulberries.

After our first meeting on a Sunday afternoon in June 1970, when Martin and Eleanor (Plate 11) immediately put me to work building a haystack, we struck up a close friendship that has continued ever since. Martin encouraged me to start designing my own sculptural clock, and over the next two years, I spent many weekends at Mulberries working under his guidance in his workshop and helping with the daily duties that go with living off the land, often hitch-hiking my way there and back to London. Since I was also working at that time with George Daniels every Thursday on completing a wooden regulator that John Harrison had begun but never finished, our conversation often drifted to Harrison's work. Could the claim he made in his 1730 manuscript for the accuracy of these wooden regulators be true? From the honest way in which Harrison described everything else he had done, it seemed to both Martin and to me that it would be unthinkable for Harrison to have lied or misled anyone, particularly about such a significant matter. After I left art school in 1972, my weekend visits to Mulberries became less frequent, but we maintained a regular correspondence. These letters, dating from 1970 to the present, have provided a very useful source of information for this article and for information on his progress on making Clock A (the Gurney regulator) and Clock B (the Saff regulator).

Following the completion of the Schroder clock, Martin continued his work on the 'hares and tortoises' clocks and in 1976, completed a sundial for the British Horological Institute at Upton Hall, but gradually his interest became more and more focused on John Harrison's regulator science. He struck up a close friendship with Bill Laycock, and encouraged him to publish his book, *The Lost Science of John 'Longitude' Harrison*, in 1976. After Bill died that same year, this publication greatly encouraged Martin and others who shared his interest in the subject to form the Harrison Research Group, whose mission was to better understand the methods and principles of precision timekeeping that John Harrison had described in his writings.

Many times, Martin and I have looked back and laughed at the number of extraordinary things that have occurred during the past forty-seven years to bring Harrison's achievements into the limelight—too many, it seems, to chalk them all up to chance. We clearly remember how underappreciated Harrison was when we first met in 1970. Although H1, H2, and H3 were admired for their mechanical ingenuity, and H4 was recognized as the remarkable timekeeper that won the longitude reward, most horologists regarded Harrison's work as 'a glorious dead end'. This attitude toward Harrison's work was widespread in the 1970s because the relative simplicity of the marine chronometer made it look completely different from H4. It seemed that that there was little recognition of the achievement of H4 and its copy K1 not only in proving that a timekeeper could be made and relied upon at sea (which immediately opened the door for others to follow), but also in their very size and basic design, which influenced the pioneers of the marine chronometer who followed in Harrison's wake.

Today, however, John Harrison has a memorial stone in the middle of Westminster Abbey next to the graves of Tompion and Graham and is widely known as one of England's great geniuses. Everyone who has been involved in the developments that have given John Harrison his due recognition could be seen as characters in a play, each carefully chosen for the part they have performed. It is a mystery—Martin and I call it the 'unseen hand'.

*****

Much of the information in this paper comes from a recording made during a lunch with Martin at the Six Bells pub in Boreham, Essex, on 15 February 2014. A follow-up recording was made at the same location four months later on 26 June. The author would like to thank the following whose insightful comments and assistance have enriched this article: Catherine Andrewes, Jonathan Betts, Bruce Chandler, Andrew King, Peggy Liversidge, Kate Mason, Rory McEvoy, Don Saff, and John Shallcross.

Unless otherwise noted, all photographs are copyright of the author.

References

Harrison, J. (1775). *A Description Concerning Such Mechanism*. London, T. Jones. Available at https://ahsoc.contentfiles.net/media/assets/file/Concerning_Such_Mechanism.pdf. Accessed 12 December 2018.

Laycock, W. (1976). *The Lost Science of John 'Longitude' Harrison*. Ashford, UK, Brant Wright.

Stevens, A. (1970). Clocks and goats' milk. *The Guardian*, 23 January.

4

Rescuing Martin Burgess's Clock B

Donald Saff

Sitting on a low window ledge of a clock repair shop on the fourth floor of a nineteenth-century New York City building was a forlorn clock, entirely patinated in dust and dirt, with many parts bent—and yet, even in that state of disrepair, it had a formidable sculptural and horological presence. Affixed to the clock plate was an undated plaque with the name 'Martin Burgess'—a name that was not familiar to me at the time. Another small plaque dedicating the clock to Dudley from Major Heathcote had been placed on the back plate in 1977. At some point, after that date, the clock was apparently put up for auction but, due to its damaged condition, it never made it to the block. Instead it was sent to the auction house basement from where it was sold directly to a collector. The repair shop owner assured me that it would likely not be restored and yet it took me three years to convince the collector to sell the clock—this, in spite of the fact that he and his colleagues also claimed it would never run. Their assessment notwithstanding, I acquired the clock and thus began my association with Martin Burgess. The clock was indeed restorable, and, once renovated, it produced somewhat consistent timekeeping for a compound pendulum. Clockmaker Richard Ketchen restored the clock between 1997 and 1999.

Around the time of the completion of the restoration of this clock, I gave a talk at the Massachusetts Institute of Technology (MIT) on experiments in art and technology in the works of artist Robert Rauschenberg. After the lecture Ketchen offered to introduce me to William Andrewes, then the Wheatland Curator of the Collection of Historical Scientific Instruments at Harvard. Andrewes, knowing my work with clocks, shepherded me through the Harvard collection, which included Richard Bond's (1827–1866) superb regulator that would capture my attention and research for the following eighteen years.

Saff, D., *Rescuing Martin Burgess's Clock B* In: *Harrison Decoded: Towards a Perfect Pendulum Clock*. Edited by Rory McEvoy and Jonathan Betts, Oxford University Press (2020). © Oxford University Press. DOI: 10.1093/oso/9780198816812.003.0004

I shared with him the recently completed restoration project of a clock by someone named Martin Burgess. Will smiled. 'I worked with Burgess,' he said, 'he is a great friend, and one of the founding members of the Harrison Research Group. You must send him pictures.' Pictures were dispatched and, after a while, a delighted Burgess replied with a 26-page, single-spaced letter, written over a period of four months. It was, and is, a truly remarkable document, as it served as my introduction to the history of the restored clock, to Burgess's work generally, and to the philosophy and the science of John Harrison.

The clock (Plate 12) was commissioned by Simon Napier-Bell, an important rock music impresario, and was meant (by his request) to mesmerise and capture the attention of those with whom he was negotiating, thereby giving him the advantage as their focus would be on the clock and not on contractual details. In a brief excerpt from a 1966 film on Napier-Bell, the cinematographer could not resist the 2½-second compound pendulum with grasshopper escapement and provided the clock with a starring role in an otherwise bland film. It worked as intended. 'Simon's Balls', as it was called, was the second of Burgess's efforts to combine sculptural horology with precision mechanics.

Also called B2, the clock was featured in a *Horological Journal* cover article (Saff, Burgess, and Ketchen, 2001). Soon after publication I received a letter from the brilliant watchmaker Derek Pratt (1938–2009), who had acquired the remaining parts from an edition of ½-size Dent replicas of the Burgess clock. Our correspondence explored Pratt's design variation of Aaron Dodd Crane's 'Daisy Wheel' motion work as the patent model for Crane's unique design, and a restored example from a Rhode Island church resided in my collection. Concerning Burgess's clock, Pratt wrote to me in 2002 stating that 'this must be one of the most desirable clocks of all time, in my opinion. It is art and science combined.' His sentiments were repeated in his March 2003 article in the *Horological Journal* titled 'More on the Double Daisy Wheel Mechanism'. Pratt donated his modified replica to the Royal Observatory, Greenwich (NMM catalogue ZBA4840).

The Burgess original was constructed in 1965, approximately 12 years before the work of the Harrison Research Group was in its incunabula stage of grasping the specifics of Harrison's science and its implications. One need only analyse its grasshopper escapement design through the parallax of today's understanding to realise that Burgess was following assertions by Cmd. Gould or Col. Quill that were misleading in representing the correct details of Harrison's design. Over the following years, beginning in the late 1970s, that would all change.

I was traveling outside the USA during the Quest for Longitude symposium at Harvard in 1993, so it was not until I met Will Andrewes and learned of Burgess's participation that I purchased the proceedings of the conference (Andrewes, 1996) and read Martin's presentation, titled *The Scandalous Neglect of Harrison's Regulator Science* (Burgess, 1996). What a title indeed—and what a remarkable read!

As a card-carrying high-Q devotee, I generally sided with high 'Q-ists' as they engaged in a tennis match of mathematical formulae in the superb *Horological Science Newsletter* and other publications (HSN, 2018). Many articles, however, were not based on empirical evidence from applied experimentation. Such pieces were in the minority. I was therefore obliged to do my own testing and scrupulously read the Burgess article again and again, as his understanding was based on years of testing. I struggled with the application of the formula for kinetic energy as applied to the pendulum and the description of the balance of environmental forces.

Burgess's (1996) chart (Fig. 4.1) had me producing my own charts for the mechanics in a given particular environment that was necessary for isochronism. This was astounding to an uninformed me still clinging to Huygens's cycloid, and to the presumed advantages of a high-Q pendulum. Just saying 'circular deviation' and not 'circular error' required an adjustment of thought. This novel approach was Harrison's high-wire balancing act of countervailing phenomena with a net result of system equilibrium and isochronism—no other net in this high-wire performance was necessary as Harrison provided and exhibited the mechanical means for its accomplishment.

Burgess opined that if there had been sufficient interest in Harrison's Late Regulator, a second in 100 days would have been a fact in the late eighteenth century and horological science would have followed his path. Imagine going the Harrison track instead of Graham's. Accuracy of a clock operating in the vicissitudes of normal air pressure that could rival clocks operating *in vacuo* could and would be a reality. I studied and became familiar with the lifting capacity of the cheeks, torque ratios, the value of recoil, the benefit of no lubrication, roller bearings, higher velocity pendulums, the quick correcting of low-mass high-amplitude pendulums with their attendant high signal-to-noise ratio, nonlinear oscillator theory, van de Pol's equations, and the 'Hill Test'.

Environmental Change		Clock Rate	Circular Deviation
Gravity	Up	+	0
	Down	−	0
Temperature up	Air thinner	Small +	−
	Rod expands	−	0
Temperature down	Air thicker	Small −	+
	Rod contracts	+	0
Energy input	Up	+	−
	Down	−	+
Barometer up	Air thicker	−	+
	Pendulum lighter	−	0
Barometer down	Air thinner	+	−
	Pendulum heavier	+	0
(With pendulum over-compensated)	Temperature up	+	−
	Temperature down	−	+

Fig. 4.1 Burgess's table showing the fundamental effects of environmental changes and circular deviation upon the rate of a pendulum. '[B]ecause of the opposing results of gain and loss, the circular deviation can be used to negate the effects of environmental changes on the rate of the clock' (Burgess, 1996:263). © Don Saff

Martin's article also introduced me to William Laycock's book, which I was eventually able to find and acquire from a local bookseller. Following other references, and especially writings by Mervyn Hobden, Peter Hastings, and Jonathan Betts, I came to the realisation that a clock could be produced and would run in normal atmosphere with superb results—only limited by constraints such as those discussed in Henry Wallman's (1992) article 'Do Variations in Gravity Mean that Harrison Approached the Limit of Pendulum Accuracy?' and in Pierre Boucheron's (1992) article on seismic activity. Though Martin's work was initially as opaque to me as reading Harrison's *A Description Concerning such Mechanism* (1775), his writing furthered my devotion to understanding both Harrison and Burgess. In the same Burgess article was a poor reproduction of a photograph of the commissioned Gurney clock (known also as Clock A) dated 1974–1987 and a reference to an unfinished sister regulator referred to as Clock B, which was also begun in the mid-1970s to assist with the process of adjusting Clock A. At the time of my reading, it remained unfinished some 25 years later.

Expressing his concern for Harrison's legacy, Burgess wrote to me in 1999, stating:

> It really has been the most frightful horological disaster, engineering disaster and retardation of the understanding of oscillator physics that clockmakers, engineers and scientists were both lazy and insufficiently humble to settle down to study the devices and testing methods invented by John Harrison. If you had died, as Harrison did, knowing that all your techniques and scientific testing had not been understood properly and that, at least as far as precision clocks were concerned would never be understood or used in time to be of any use to science, do you think you would have 'rested easy in your grave?' I don't think so. I have always felt that I was working for him rather than myself.

Though an art historian by profession, I was always devoted to an understanding of electronics and was licensed by the Federal Communication Commission as a First Class Radiotelephone Operator with Radar Rating. Burgess's reference to oscillators touched on a field of physics with which I was quite familiar and further piqued my interest.

Perhaps the Royal Astronomical Society regulator would have been finished if there had been sufficient interest and had Harrison lived long enough. Now, why was the second Burgess regulator not finished and, using Burgess's own words, would it not be *The Scandalous Neglect of Burgess's Work* in demonstrating Harrison's science if Clock B was not completed and tested? Burgess was not making a reproduction but an example of Harrison's general theory and science to prove Harrison correct while using modern materials.

I found a 1989 article on the Gurney clock (Clock A). There was no mention of the unfinished regulator but there was a reasonably good photo of the Gurney clock (Tyler, 1989)—once again I was struck by the beauty of Burgess's design. The images reinforced my appreciation of the simplicity and elegance of Burgess's clock design as applied to Harrisonian horology.

I immediately called Will Andrewes and asked whether he would help me bring the completion of Clock B to fruition. And, if anyone knows Will, then you know he was indefatigable in his efforts to convince Martin that it would be finished to the highest standard and adjusted and tested to prove, or disprove, the efficacy of Harrison's science—this was done with the advocacy of Andrew King. Other luminaries agreed to consult so there could be additional informed input to the process.

Will and I visited Martin's Schroder Bank Clock (Plate 13) as I gathered information about Martin's mechanical and stylistic proclivities. From Canary Wharf, the new home of the bank clock, we went on to Norwich in June 2009 to see the Gurney Clock (Plate 14), which was located in a lonely part of a shopping centre displayed in a smartly built circular kiosk. What an improbable setting. Yet for me, it was like seeing a Northern Renaissance painting. First, one sees the overall beauty and then comes the recognition of exquisite detail next to exquisite detail. From Norwich, Will and I went on to Boreham, which was resplendent in colour as it was celebrating the Flower Festival. I wanted to see the church and while I was checking the architecture, Will knelt and said a prayer. Then on to visit Martin (Plate 15).

We considered many production plans and horologists for completing the project but in the end, all roads seemed to lead us to Charles Frodsham & Co. Directors Philip Whyte and Richard Stenning were invited to our meeting with Burgess that day. Initially, I was unsure of the way forward but as I strolled back separately with Philip from a lunch with Martin, his wife Eleanor, Andrew King, Richard and Will, Philip said: 'I don't know where this project will lead but we will do our best and we will be fair.' I had concluded that if all agreed, Frodsham was best suited for the challenge. Martin approved, and Philip and Richard agreed to participate, utilising their gifted horologists/artisans—Roger Stevenson and Martin Dorsch. That gave us access to the necessary expertise, skills, and equipment yet none of us knew exactly where we were going, how and if we could get there, and what the results would be.

Will and I returned to Boreham with Philip, Richard, and Andrew King. The unfinished Clock B was installed in the same location on the second floor of Martin's shop as the Gurney clock when it was still in production. The great wheel of B was hanging near a lower level window; the lathe fixture illustrated in *The Quest for Longitude* (Andrewes, 1996) that Laycock had designed for accurately cutting the cheeks was in a cardboard box; and the escapement was housed in a separate and secure wooden box that was adorned with formidable warnings.

The clock was removed and brought to Frodsham's facility in Punnetts Town, East Sussex. The finish, construction, and testing of Clock B in their workshop are described in the next chapter. Frodsham's liaison and consultation with Burgess, and their preclusion of my micro-managing, yielded the superb results that are fully described in this volume. I, for example, wanted a differential barrel for winding the clock but the team deemed it too cumbersome and were fearful that if not regularly wound the test would be compromised. Instead, they used Martin Dorsch's innovative electric motor configuration which combined constant torque and dependability. There were those whose scepticism only abated when the initial Frodsham trial exhibited unprecedented results, and it was apparent that the clock kept time to almost within a second in 100 days, under less than ideal conditions.

On 28 June 2011 we convened a meeting in Punnetts Town to discuss any remaining details to be resolved in completing the clock and the way forward to trialling the clock with rigorous integrity and transparency. At that meeting was Jonathan Betts, then Senior Horological Specialist at the National Maritime Museum; David Thompson, then curator of Horology at the British Museum; Will Andrewes; Andrew King; Philip Whyte; Richard Stenning; Roger Stevenson; Martin Dorsch; Daniela Hofer; Ruth Saff; and myself (Plate 16). We agreed that the formal trial would be held at the Royal Observatory Greenwich under Betts's stewardship after the clock was completed and that Jonathan would make the final adjustments.

The Frodshams team delivered the clock on a rainy day, 18 April 2012, to the Royal Observatory and installed it on the base support for the Great Equatorial Telescope. Thirty-eight years had passed since Martin started its construction. Testing was then under the supervision and direction of Jonathan Betts assisted by Rory McEvoy. Jonathan made the necessary adjustments and modifications, which included the Hill Test and the change to the suspension spring to achieve the correct position of the running arc on the hill. Rory tracked the rate, reporting regularly to all the participants. As testing continued, more responsibility was carried by Rory McEvoy, who resolved issues with the testing equipment and further trialling of the clock.

The formal trial of Clock B in 2015 was witnessed and attested to by peer reviewers who would act as overseers, thereby guaranteeing the integrity of the trial. Amongst other organisations represented were the National Physics Laboratory and the Company of Worshipful Clockmakers. The results of the trial were memorialised by Guinness World Records on a certificate (Plate 17) which states: 'At the beginning of a 100-day trial "Clock B" was ¼ seconds behind Coordinated Universal Time (UTC); it ended the trial 7/8 seconds behind UTC as verified on 16

April 2015.' Without the application of any rate, the clock remained within 5/8 of a second in 100 days.

Scrutiny of the clock's performance revealed that adjustment to the suspension spring (to improve barometric compensation) had created a new and unexpected reaction to changes in temperature. The fixed brass tube compensator was now under-compensating. In late 2017, Rory constructed telescoping brass tubes of Burgess's design with 100 threads per inch for fine adjustment, readying the clock for continued testing.

Over the course of numerous trials, so closely was the regulator observed that in my daily rush to the computer to see the rate each morning I fancifully thought that we could be leaving classical mechanics and entering into the realm of Heisenberg's quantum physics and observation error. Silly as that may be, it was the result of dedicated research and consultation that morphed into near obsession as the clock's unprecedented stability became ever more apparent. The analyses and data in this volume will always spark controversy, but the insights provided are unimpeachable evidence that Harrison's claims had scientific veracity and that Burgess's understanding of those claims has been truthfully exhibited.

References

Andrewes, W., ed. (1996). *The Quest for Longitude: The Proceedings of the Longitude Symposium Harvard University, Cambridge, Massachusetts November 4–6, 1993.* Harvard, Collection of Historical Scientific Instruments.

Boucheron, P. (1992). Of earthquakes and clocks, *Horological Journal*, July, 26–28.

Burgess, M. (1996). The scandalous neglect of Harrison's regulator science. In: Andrewes, W., ed., *The Quest for Longitude: The Proceedings of the Longitude Symposium, Harvard University, Cambridge Massachusetts, November 4–6, 1993*, pp. 256–278. Harvard, Collection of Historical Scientific Instruments.

Horological Science Newsletter (2018). Available at http://www.hsn161.com/HSN/newsletter.html. Accessed 5 December 2018.

Saff, D., Burgess, M., and Ketchen, R. (2001). Burgess revisited: new life for 'B2'. *Horological Journal*, August, 268–270.

Tyler, E. J. (1989). The Gurney clock—a mini guide. *NAWCC Bulletin*, June, 236–237.

Wallman, H. (1992). Do variations ion gravity mean that Harrison approached the limit of pendulum accuracy? *Horological Journal*, July, 24–26.

5

Reflections on Making Clocks Harrison's Way

Martin Burgess FSA, FBHI

The visual design of a precision clock has nothing to do with its accuracy, nor has a beautiful finish, so long as everything is right in the right places. Harrison's Late Regulator looks nothing like Clock B but would have been just as accurate, if not more so, had it been completed in the mid-eighteenth century.

Anyone setting out to make a precision Harrison pendulum regulator must follow specific rules. These are not based on higher maths, but on logic and experiment, or as Harrison says, 'reason and experience'. It was a method well advanced by him by 1730 but then perfected and expanded until he died in 1776. This methodology had not been effectively applied to clockmaking since Harrison's death, until the formation in the mid-1970s of the Harrison Research Group, which started to unpick Harrison's methods from the surviving documents and manuscripts.

It is worth noting that Harrison did not develop his model of precision clock entirely on his own. He would have been in touch with all the latest thinking concerning the mathematics and physics of oscillators after he moved to London, certainly through George Graham if no one else.

Prerequisites for a Harrisonian precision pendulum clock

1 Oil must not be used anywhere in the mechanism. Harrison used brass pivots in lignum vitae bushes in his earlier precision clocks. His first oil-free clock was the turret clock over the stables at Brocklesby Park (once he had replaced the anchor escapement

Burgess, M., *Reflections on Making Clocks Harrison's Way* In: *Harrison Decoded: Towards a Perfect Pendulum Clock*. Edited by Rory McEvoy and Jonathan Betts, Oxford University Press (2020).
© Oxford University Press. DOI: 10.1093/oso/9780198816812.003.0005

with the newly invented grasshopper escapement). It has been running now for almost 300 years and there is no wear on any of the pivots. Oil changes its lubricity with temperature. It also evaporates, spreads, ages, and collects dirt. Harrison expected his regulators to go for at least 100 years without stopping, cleaning, or wear, provided all dust could be kept away from it.

2 The wheel train needs to have very low friction. Harrison used roller pinions. The rollers are lignum vitae on brass pins, and to get large rollers he had to use narrow, straight-sided peg teeth. Clock B uses stainless steel dry Barden bearings. They are dust shielded, and the cages are coated with Teflon, which means they offer very little friction. Furthermore, they are small enough to engage with cycloidal teeth, thereby enabling a reasonably constant torque transmission. A low-friction hard plastic might be a useful material for the rollers. In the Late Regulator the first pinion, driven by the great wheel, has a high number of rollers to make the drive smoother. According to Laycock, Harrison said there is no engaging friction in recoil because the rollers only 'rocked upon their pins, due to the running clearance' (Laycock, 1976:38).

3 It is essential to have a constant energy input to the pendulum, and it will need to be large, which is good, for Harrison says 'a large force changes in proportion to itself less than a small one'. The train must end with a well-designed remontoire that does not abstract much energy from the escape wheel during the unlocking process. To achieve this aim, the unlocking action also needs to be spread out over almost all the half-minute. Improving on the unlocking mechanism in Harrison's remontoire is difficult. His escape wheel turns once in four minutes, and so requires a four-bladed second hand.

The remontoire design

The unlocking function in Clock B consists of a pivoted arm, at one end of which is a swinging claw that engages with a ring of eight pins on the escape wheel arbor. The claw engages with a pin, which is at the tangential point, and is lifted and drawn inwards as the pin ring turns. This motion goes on until the train is unlocked. A cam on the fly arbor lifts the detent arm. The claw disengages and swings outwards to its stop, almost touching the next pin, which is now in the tangential position. It is essential that the claw does not fall onto the next pin as the remontoire locks, for if it did so, the shock might discharge the engaged pallet of the escapement.

The detent pallet itself is near the centre of the arm's motion thus de-multiplying the lifting energy. The whole unit needs to be counterpoised so the energy required for lifting is minimal.

The detent arm, when down, has the detent engaging with a blade that is fixed to the fly arbor. The contacting parts are tangential to each other so avoiding any chance of giving motion to the arm when the train is locked. This arbor, at the end of the train, also carries the cam to lift the detent arm and return it. As with Harrison's, my detent arm has a small anti-friction wheel that rests on the cam. The fly is friction spring-loaded on its arbor so, when the detent arrests the motion of the fly arbor, the fly can go on turning a little way, absorbing the momentum.

The two helical remontoire springs, found on Harrison's third sea clock, have their inner ends attached to metal ribbons acting on cams. This arrangement strongly suggests that he wanted the remontoire to deliver precisely equal torque to the escape wheel regardless of the state of wind, whether fully wound, half-wound, or about to be wound. In this way, changes in the state of the air and the speed of the fly should not change the energy imparted to the pendulum. The single spiral remontoire spring, found on Clock B, is of Ni-Span-C, which has a very low thermoelastic coefficient.

The fly on Clock B only makes one turn at rewind and has four short heavy blades, so it works almost like an inertial fly. The design might be improved by employing an inertial fly: two small, streamlined weights at each end of a rod pivoted on the fly arbor. In this manner there would be less change of speed under different air conditions, it would take up much less space, and it would have been quicker to make. Harrison's fly would be much more sensitive to changes in atmospheric pressure and temperature.

The grasshopper holds the key to the whole of this pendulum technology. The escapement on Clocks A and B follow Harrison's original design and were constructed as per his accurate working drawing (held in the library of the Clockmakers' Company). The dimensions are the same as the surviving parts of the Late Regulator; the pallets and composers on that are non-original.

When the pendulum is vertical and static the pallets span 16.5 tooth spaces. The escaping arc is about 10° and it appears from the texts that Harrison was running at an amplitude in the region of 12° or 12.5°. For the testing, the clock must be able to run at 13° or more.

The repeated energy coming from the escapement to the pendulum is quite complicated but can be described. The escapement does not require oil and is almost free from friction and wear. This quality allows the escapement to run for many years without any change to its characteristic mode of operation.

Let us say that the pendulum is moving from right to left. When it reaches the escaping arc, out of the recoil part of the arc, the thrust on the pallet, tangential to the rim of the wheel, is as two units (taking as an average between the pallets). As the pendulum moves towards the left the thrust rises until, at the opposite escaping arc, it is as three units. The interchange now takes place and the thrust instantly drops to two units. It then goes into recoil, the thrust on the pallet dropping under two units, more and more until the running arc is reached, 12° to 12.5°. The pendulum then returns, the thrust growing to two units at the escaping arc of about 10°. On and on, repeated and repeated.

The crutch arbor needs to have very low friction. On Clock B and Harrison's clocks this arbor has knife edges at each end. Harrison's pendulum clocks use glass, whereas Clock B uses sapphire v-blocks. Energy from the escapement is communicated to the pendulum by the crutch—an armature that is fixed to the back of the pallet arbor. Harrison would have probably used brass for the crutch. As the crutch swings with the pendulum, it must be considered a part of the oscillator. I opted not to use brass, following the advice of Bill Laycock, who suggested that the copper and zinc alloy may not have a stable coefficient of expansion in heat in the long term. Clock B uses an Invar crutch with a stainless steel fork.

The Harrison Pendulum

The pendulum is undoubtedly the most critical unit in the clock system since it is the timekeeper. Careful thought must be applied to its design. Harrison's gridiron is entirely different from any other type of pendulum, and its function goes beyond mere temperature compensation.

The pendulums on Clocks A and B lack air resistance since when they were made, I had not fully understood the principles required and

was still under the influence of traditional clockmaking. Two statements by Harrison are important and must be held in mind all the time:
A large force varies in proportion to itself less than a small one so
energy-efficient pendulums are to be avoided; and 'the air's resistance
[does not want to be avoided] as some people have so foolishly imagined
but is of real or great use' (Harrison, 1775:27).

A pendulum that consumes a lot of energy requires significant
energy input. Disturbances from the outside transmitted through the
wall to the pendulum such as traffic in the street or the slamming of
doors (both mentioned by Harrison) must be swamped. The pendulum's stored energy equals half the mass times the velocity squared, so
we need a broad arc of swing to achieve high velocity. Clockmakers
have traditionally avoided large arcs of oscillation because they were
keen to minimise the timekeeping errors that arise from a change in
amplitude. At 12°, circular deviation (some call it circular error) is substantial even for a minimal change of arc. Harrison used circular cheeks
to speed up the longer arcs.

In his wooden regulators the cheeks are not tight on the suspension
springs but are spring-loaded and linked together by a pivot; a single
adjustment nut controls their positional relationship. The best position
is found through experiment. Different arcs are generated by varying
the energy input to the escapement. In the 1730 manuscript, he described
the use of different driving weights. However, in the Late Regulator, the
cheeks are not adjustable and therefore must be the correct radius
to begin with. This required radius will vary accordingly for different
pendulums and must be found for each clock by experiment. It is not
an easy matter to solve.

The pendulum suspension unit with its cheeks is not only vital to
the Harrison technology but is extremely sensitive to any dimensional
changes—whatever the cause. The cheeks must be precisely the right
radius, and the suspension spring must be just the right stiffness.
Harrison (1775:45) says, it must be 'thin to the purpose'.

Harrison tells us, in the 1730 manuscript, that later he found fixing
the suspension unit to the clock case to be wholly unsatisfactory. He
drove two strong irons into the joints in the brickwork of his house and
fixed the suspension unit to those. They must have stuck out from the
wall to make room for the clock case because there are surviving holes
to let them through. In this way, Harrison insulated the suspension

unit from the influence of the wall. It needs to have the air temperature of the room and not of the wall behind.

I had a bad experience of fixing brass cheeks to a steel block bolted to the 0.5"-thick steel plate which is the backboard of Clocks A and B. One of the two clocks was on an external cavity wall, and this showed marked changes of timekeeping over day and night. This caused me to look again at the suspension unit on the Late Regulator only to find that Harrison had been there before me.

Temperature error due to the suspension unit must be compensated in the pendulum. In the Late Regulator the brass plate that supports the cheeks is quite long, as are the cheeks. Both are close to the same volumes. The attachment to what is behind is right on the ends, well away from the middle. The outer ends of the cheeks have a large hole in them to let through a brass pillar attached to the cheek back plate. These pillars locate the movement of the clock but they are not fixed to the back plate of the movement. Instead they pass through it and are fixed to the front plate so that they can be as long as possible.

It is apparent that these pillars served as heat-feeders, which ensured that the suspension block was influenced by the air temperature of the room and not of the wall behind. We do not know what insulation Harrison intended behind the unit to protect it from the wall. He complained that he had nowhere suitable to fix it in his house.

He was going to give it to Greenwich Observatory and said it must be installed in a pretty temperate place. This would probably be the basement of the observatory. He also says that nothing must be in contact with anything made of wood (lignum vitae being the exception of course). Wood changes its dimensions a lot in different humidities. Looking at the Late Regulator, we can see that the movement plates are much broader at the bottom and the fixtures for it are right on the outside, so the movement is bridge-like. This design informs my conjecture that Harrison intended to mount the clock in a brick alcove with the insulated suspension unit fixed to the back wall and the movement bolted down to small stone ledges sticking inwards from the side walls. Then the whole thing would be glazed in and made to be entirely dust-proof. Air in and out would have to pass through a very good filter. Harrison expected his regulator to run for at least 100 years without wear or stopping and to be within two or three seconds a year, provided all dust could be kept away from it.

Realising that the suspension unit must have the temperature of the room and not of the wall behind, I remade the support block the same thickness as the cheeks and screwed it on to two 303 stainless steel pillars, which were fixed into the steel backboard of the clock. A slotted barrier of wood was placed over the pillars to stop radiation between the wall and the suspension unit. This arrangement undoubtedly improved the situation. However, it was not possible at the time to quantify the improvement.

The gap between the cheeks has the suspension spring in it, of course. In Clocks A and B, it is a 0.5"-wide ribbon of Ni-Span-C. The gap between the cheeks is critical. So, to get it always the same, the cheeks are pushed onto the spring, which is about 0.005" thick.

Finding the correct radius and suspension thickness requires patient experimentation. These are likely to involve many months of careful adjusting, observing, and testing. How long it takes is of no importance. It is perhaps advisable that anyone building such a clock should, like me, replicate the cheeks of the Late Regulator. Clocks A and B use a radius of 3.538", as measured by Bill Laycock.

The hill tests

The dichotomy of Harrison's clock system and the traditional model of pendulum clock is profound. Harrison's system uses an anisochronous pendulum, and this is necessary to compensate for changes in air density (caused by both barometric pressure change and temperature change).

On the rim of the remontoire wheel on H3, there are two cams, each with five positions, so that each remontoire spring can be stretched to exert five different levels of torque. The Late Regulator has a rim on the remontoire wheel with identical provision for two cams. These allow the clock to be timed at five different arcs, two before the running arc and two after it. On Clocks A and B, different setups of the remontoire are used to get a range of arcs.

By activating the remontoire six times (from an entirely unwound state) the running arc of a little over 12° is achieved. The clocks run on five setups of the remontoire, but four setups reduce the arc to a point where the grasshopper can no longer escape. The initial testing, required to adjust the clock's air density compensation, was conducted over five, six, and seven setups. Using only three points on the graph was not ideal, but was just enough to show what was going on.

The horizontal scale of the graph showed the arc of swing, say $10°$ to $14°$, and the vertical index showed the rate in seconds per day. Once the suspension spring and cheeks are well matched, the points will delineate part of a bell curve. This curve was termed the 'hill'. The aim was to place the running arc on the far side of the 'hill' with exactly the right circular deviation in different pressures to cancel out the air density error.

The clock was regulated at six setups, and then the rate and the arc of swing was recorded at five and seven setups so that these arcs could be returned to again and again without having to wait long for the arc to settle down. The pendulum does settle down very quickly because of the high air friction and the large quantity of energy passing through it. However, this process can be sped up by increasing or reducing the arc by pressing on the crutch with a springy hook-shaped wire. In this way, measurement for each setup took around ten minutes, using a Bateman Radio Check-rate. There are more convenient instruments for this now.

Initial results showed a steep line, the clock running much faster in the wider arcs. Then I withdrew the cheeks in their cradle on the face plate of the lathe by 0.005" and turned the curve back to that. I repeated this many times. I would have done better to go straight to Bill Laycock's measurement of the radius on the Late Regulator cheeks, 3.538".

We don't know what Harrison would have used for the suspension spring. It was unlikely to have been steel but it might have been well-hammered brass like on Brocklesby or, on the Late Regulator, it might have been high-copper gold.

By gradually reducing the radius of the cheeks, the clock's behaviour changed. It did not run so fast in the broader arcs, and the 'hill' began to appear on the graph. The rate was slower in the shorter arcs and also slower in the longer arcs. This pattern was most unexpected, and so I telephoned Mervyn Hobden, and asked, 'Should I see a hill coming towards me?' He told me that this was exactly what I should see, so from then on we called these 'hill tests'. From the evidence found in H3 with its five-position cam, we can be reasonably sure that Harrison used the five different repeatable arcs to fine-tune the compensation.

The design of the pendulums for Clocks A and B

This pendulum works but leaves much to be desired. It has not got enough air resistance. The temperature compensator is not readily and finely adjustable without taking the pendulum off the clock. An improved adjustable compensator has since been made and fitted by Rory McEvoy, made to my own design.

To compensate for the contraction and expansion of the rod, most clockmakers support the bob on a brass tubular compensator inside the middle of the bob. But hiding the compensator inside the bob renders the compensator almost useless because its temperature will not change until the whole bob changed. By which time, the rod's length may have changed long ago, and the temperature may have reverted.

In Clocks A and B, I made the lenticular bob in two halves with a gap between them and bridges to unite them. The lower bridge is in the centre of the bob and is supported by a compensating tube whose length was obtained by calculation. This is quite wrong, for it has to compensate for all changes of time caused by temperature change, including changes of arc caused by the changing state of the air.

The air resistance of these pendulums is a little too low and needs to be made adjustable. Changing air density affects the arc of swing and adjusting the pendulum's resistance would enable changing the shape of the 'hill'. The running arc needs to be on the downward slope on the far side of the 'hill'. This desirable position on the graph illustrates that there is a bit of necessary circular deviation left over at that big arc of 12° to 12.5°.

These pendulums could have adjustable air resistance in the form of two little vanes back and front, about the size of postage stamps. These might be of very thin brass silver soldered each to a half-tube embracing the pendulum rod, made so that it can be clamped in place. I would start with it two-thirds of the way down from the suspension, further down to increase the effect and up to reduce it. Harrison would have got quite a lot of air resistance because of its being a gridiron pendulum.

Air density compensation

In changed barometric pressure or air temperature, my pendulum, as it is now, hardly changes its arc at all, so the far side of the 'hill' needs to be very steep to remove the air density error. Harrison is brutally frank about this part of the subject without explaining it: 'the air's resistance does not want to be avoided as some people have so foolishly imagined but is of real or great use'.

The error is due to the changing air density or, as Harrison said, the weight of the air. In higher density, the weight of the air that the pendulum displaces is more, just as when we get into a swimming bath we are lighter—floating in the denser medium, water. The change in timekeeping will depend on the volume of the pendulum, but this floatation effect lessens the effect of gravity acting on the pendulum and the clock runs slow. At the same time the greater density of the air makes the pendulum decrease its arc and circular deviation will cause an increased rate because of the reduced arc. However, this is a much smaller effect on the timekeeping, though it is tremendously important to the Harrison system of compensation. Returning to the graph, or 'hill', on an adjusted clock the running arc point will move back up the far side of the 'hill'. These two things must be brought together so that they cancel each other.

Further adjustment to the gradient of the 'hill' is possible through the escapement design. In recoil the entry pallet pushes the escape wheel backwards and the exit pallet pulls it backwards. In doing so, both lift their composers off their stops. In this way, the composer's weight speeds up the longer arcs and thus makes the slope of the far side of the 'hill' gentler. In recoil, the remontoire spring in clocks A and B are also wound up slightly and this too will change the shape of the far side of the 'hill'. On the Late Regulator the existing composers look very heavy and this may be because the gridiron has a lot of air friction and causes a big change of arc under different conditions.

Jonathan Betts at Greenwich managed to achieve this, in spite of the low air resistance of the pendulum. The running arc for Clock B must be on a very steep 'hill', for in very great changes of pressure the arc can be seen to change very little. Only with an isochronous pendulum will the air density error show itself in full; otherwise, there will be circular deviation mixed up with it and may totally swamp it.

Temperature compensation

It's important that the temperature compensation be easily adjustable without taking the pendulum off the clock. It can only be made correct by experimenting at different temperatures. Harrison's description in the 1730 manuscript is fascinating but is out of place in this paper. He reported that he could see a change of time to a twentieth of a second or better but with our instruments we can see absolutely microscopic changes of rate.

For an Invar seconds pendulum a compensator of brass about 1" long is about right for a normal pendulum bob, but for the light bob needed for a Harrison pendulum (not more than 3 lb) it will have to be moved further to compensate and the compensator will have to be much longer as a result. However tested and perfected it is, it will still be wrong. After correcting for the air density change, caused by changes in barometric pressure and temperature, the temperature compensation will have to be adjusted again because a change of temperature also causes a change in air viscosity, which tends to act against the other correction. It may be that both errors cannot be brought into a perfect balance. Harrison tells us to keep the clock 'in a pretty temperate place'. He also says the pendulum needs to be 'shorter when warmer'.

For a gridiron pendulum Harrison's invention of the tin whistle fine temperature compensator is superb. I have been told that some clockmakers made tin whistles but put them inside the bob. This was a really stupid thing to do because a fine adjustment needs to be exposed to the air and easy to get at. Britten's handbook suggests a compensator supporting two cylinders (probably brass cylinders filled with lead) for the bob with bridges between them. The bob should not weigh more than 3 lb and should not be lenticular, so, for my clock, the compensator should be in two parts—two tubes screwed together with very fine threads. The inner tube, if made of brass, might be 1.75" long. Its top would support the bridge and just below that a square of brass plate silver soldered to it, a snug fit between the halves of the bob. This is so it cannot turn on the rod. The outer and lower tube is threaded inside and at its bottom has a ring of tommy bar holes so that a probe can be inserted in one of them to screw up or unscrew the outer tube on the inner tube, thus changing the length of the compensator.

If the rod is Invar, the rating nut also needs to be of Invar and graduated. It might be necessary to hold the outer compensator tube with a probe so that its position is not changed. For a clock as accurate as this, adjusting the rate by the rating nut is far too coarse. Harrison talks about 'the nut at the bottom'. I think he refers to a small additional nut on an extension of the thread on the rod below the bob. My clock has a small brass nut there. One turn makes a difference of about 0.1 second a day so if it was divided into five parts, half a part would be 0.01 seconds per day. The use of a little nut like this was well known in Harrison's day.

Summary

Regardless of the design of the regulator, certain rules have to be followed and experiments made. Materials must be chosen which do not need lubrication, are low friction, and are resistant to wear. *No oil!*

There are four key components that must be there, and must be right: the pendulum, the suspension, the escapement, and the remontoire.

The pendulum is served by the others and must be made correctly by experiment. It must have a high stored energy but stored as velocity and not as mass. The bob must not weigh more than 3 lb. It should have a large energy throughput so plenty of air resistance and a large arc (over 12°) so that it throws off outside interference quickly. It must have a finely adjustable temperature compensation that can be changed without taking the pendulum off the clock. The compensator must be fully exposed to the air. If a gridiron pendulum is used, the rods must be little more than thick wires to change temperature quickly, not lacquered, and the brass thicker in proportion so both brass and steel reach the new temperature at the same time.

The suspension unit is critical to the whole technology. It must be rigidly fixed to a solid wall, yet not influenced by the temperature of that wall. The radius of the circular arc cheeks must be correct, probably within.001". The suspension spring must be exactly the right thickness. Each pendulum will be different and the suspension must remove almost all the large circular deviation at that big arc, leaving just enough to cancel the error caused by air density change. The correct radius of the cheeks and the suspension spring thickness can only be found by repeated experiments.

Make the vital grasshopper escapement according to the exact dimensions of Harrison's original working drawing, which cannot be improved on. He used brass for everything except for the pallets, which would have been oak. I used duralumin throughout but my pallets are faced with lignum vitae, which was probably not necessary.

The remontoire unit can be of any design to unlock every half-minute. It must supply the escape wheel with a constant energy input indefinitely. The unlocking of it abstracts energy from the escape wheel and should be spread out over almost all the half-minute. The time taken for the remontoire to rewind must be absolutely constant regardless of the environmental conditions.

Reference

Harrison, J. (1775). *A Description Concerning such Mechanism*. London, Jones, T. https:// ahsoc.contentfiles.net/media/assets/file/Concerning_Such_Mechanism.pdf. Accessed 12 December 2018.

Laycock, W. (1976). *The Lost Science of John 'Longitude' Harrison*. Ashford, Brant Wright Associates.

6

Completing Clock B

Charles Frodsham & Co.

Delivery and initial plans

Clock B was collected from Martin Burgess at Boreham on 6 June 2009 and was delivered to the Frodsham workshops that evening. The steel back plate, weighing approximately 90 kg, was a particular challenge to get up the stairs! Having hung dormant in Martin's workshop for around twenty years, the clock looked a little tired (Plate 18), so we cleaned the movement and gleaned information on both Clocks A and B. By attaching a weight and line to the second wheel, which carries the minute hand, it was possible to see how Clock B ran before making any alterations. The great wheel had been cut but not mounted on its arbor and the second wheel needed the lantern pinion, ball races, and a means of mounting the minute hand.

Discussions with Martin and Mervyn Hobden about how best to complete the clock were had and their recommended improvements, in the light of their experience with Clock A, were incorporated, along with considerations on the desired aesthetics, to retain a strong family resemblance in the finish.

What could be expected from Clock B?

There were many anecdotes about the going of Clock A, but it wasn't clear how well it had run. Hopes were high that Clock B would perform well, but some reassurance that Clock A had shown sufficient promise was needed.

Martin had given a lecture at the Longitude Symposium at Harvard in 1993, during which he presented a graph of Clock A's rate, which showed that the maximum error was less than plus or minus one second over a

Frodsham, C., *Completing Clock B* In: *Harrison Decoded: Towards a Perfect Pendulum Clock*. Edited by Rory McEvoy and Jonathan Betts, Oxford University Press (2020). © Oxford University Press. DOI: 10.1093/oso/9780198816812.003.0006

period of 46 days (Burgess, 1996:272). This data was taken during a test
undertaken in 1986. The clock then went to Norwich and then to the
British Horological Institute in 1993. Unfortunately, it was not possible
to track down any record of the clock's performance over this period.

Clock A

The weight drives the train of wheels to the remontoire and thence to
the escapement at the top of the clock. The lowest and largest wheel, or
great wheel, carries the hour hand that rotates once in twenty-four
hours; the next wheel up, or second wheel, rotates once per hour and
carries the minute hand; and the top, or escape wheel (through the
remontoire), rotates once in four minutes and carries the seconds hand.
The unusual feature on the Burgess clocks is the remontoire, which is
placed between the second wheel and escape wheel and is designed to
isolate the escape wheel from any variations of torque resulting from
errors in the train. This is achieved through the winding of a flat spiral
spring every thirty seconds and this spring drives the escape wheel
directly. The escapement has large recoil, but the remontoire spring
absorbs the reverse motion and isolates the whole train.

Visiting Norwich to take photos and measurements also provided
an opportunity to speak to Richard Price-Thomas, who had been
looking after Clock A since it was installed at the Castle Mall
Shopping Centre in 1999. Richard spoke highly of the clock's close
rate and reliability.

Carefully rating a clock is a time-consuming business and Richard's
observation of the clock was mainly limited to changing the time indicated from GMT to British Summer Time. Over the ten-year period
that the clock had been running at the mall, the only snag was a build-
up of a sticky substance on the remontoire fly pinion rollers. The clock
was adjacent to the shopping centre's car park and a tar-like deposit was
responsible for the sticking. Richard cleaned the rollers and the display
case was made more dustproof.

A Microset timer was then connected to Clock A in Norwich so that
it could be remotely monitored at the Frodsham workshops in Sussex.
In addition to this, Richard visited the clock and reported the time
indicated. From 24 November 2010 to 25 January 2011 the clock lost
2.0 seconds over 62 days. A further check on 10 June 2011 showed the
clock had lost 16.5 seconds over a total of 198 days, and the clock had

been running for ten years. Not one second in a hundred days, but encouragement enough to persevere with Clock B.

Starting the work

In spite of Martin's rather intimidating warning (Plate 19), work was started on the escapement! During an earlier test of Clock A in 1982 the clock had run well until one of the composers stuck, causing a significant change of rate. The composers and pallets were pivoted on ball races and Martin and Mervyn changed the bearings in Clock A to lignum vitae bushes bearing on a gold pin. Clock B had its original ball races and these were changed to bushes made from Polyetheretherketone (PEEK). PEEK has a low coefficient of friction similar to those of Teflon and lignum vitae, but is stronger and harder and turns nicely.

These PEEK bushes now bear on a gold pin. The remontoire detent claw was also pivoted on a ball race and this was changed to a PEEK bushing that also bears on a gold pivot. The pallet arbor, which is one with the crutch, was also originally pivoted in ball races (Plate 20).

These had been changed on Clock A to a pallet arbor pivoting on brass knife edges that rest in a V-shaped notch in a sapphire pad. Anthony Randall had made the brass pallet arbor and notched sapphire pads for Clock A and sent us spare pieces of sapphire for Clock B. These needed to have the notch ground and polished to work with the new pallet arbor (Fig. 6.1). The arbor was made in two parts, which is necessary for assembling the pallet frame onto the arbor. The part was turned and milled in a Schaublin 70 lathe.

The sapphire was ground using V-shaped copper laps charged with diamond paste working to finer grades to produce a polished finish. An old 6-mm watchmaker's lathe was used for this work. The laps were held in the headstock and the sapphire blank was held on a vertical slide. The final polishing was done by hand using copper polishers.

It was essential to have the knife edges on the centre line of the pallet arbor, and this was achieved by first grinding the sapphire and polishing the notch, then measuring the position of the notch in relation to the rectangular sapphire block. A brass holder was then accurately milled for the sapphire, ensuring that the notch was at the centre of the brass bushing. This brass bush containing the sapphire notch replaced the ball races for the pallet arbor. Clock A has two ruby endpieces that limit the pallet arbor end-shake, but on Clock B we fitted a sapphire window at each end.

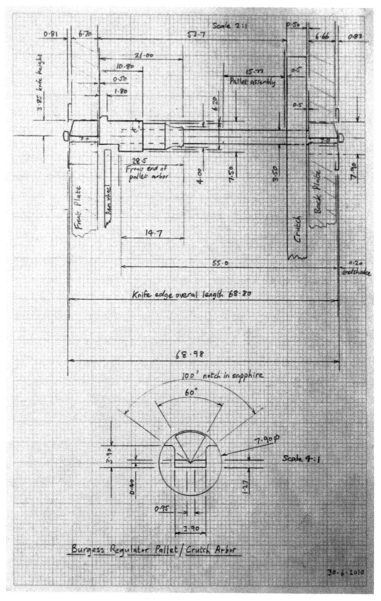

Fig. 6.1 2010 design drawing for the replacement pallet/crutch arbor.
© Charles Frodsham & Co

All the train and escapement pivots run in ball races, and Martin used ball races in place of lignum vitae rollers as used by Harrison. Ball races with a 3/16" outside diameter were purchased for the second wheel, remontoire wheel, and fly. These stainless steel ball races are shielded to keep out dust, and they run dry. Apart from the rewind motor, the clock does not require oil. This condition must contribute to the clock's long-term stability and very low maintenance requirements.

Construction of the electric rewind and work towards the finishing of Clock B

As in Clock A, it was decided that Clock B should also be equipped with a system of electrical winding. The system employed by Martin on Clock A was an endless Huygens's-type weight drive, using a link chain and sprocket pulleys. Input and output pulleys are arranged concentrically on the great wheel arbor. During winding, the input pulley is rotated using a ratchet and pushrod system, which allows the motor to be hidden underneath the floorboards. As far as we know this system only needed attention once in twenty-four years—an impressive achievement!

Various suggestions for the rewind to be fitted to Clock B were discussed, one of them being the differential system Martin had used on the Schroder Clock. This design would have required a large mass, so the idea was discarded. The next option considered was a simplified endless roller chain drive with a single loop (Fig. 6.2). It would employ a weight with an inbuilt motor, climbing up the chain once it had reached a certain dropping point. The power feed to the weight/motor unit posed an aesthetic challenge and, for that reason, a different system was ultimately chosen. It consists of a weighted arm, mounted on a unit that connects to the great wheel arbor (Fig. 6.3). Again, there is no need for maintaining power. The device is slim enough to be mounted behind the great wheel bridge and in front of the pendulum, therefore rendering the rewind system quite unobtrusive.

A number of advantages can be achieved with this design. The view of the pendulum and its energetic action remains unobstructed. The amount of driving force is easily adjustable even with the clock running, either by changing the lever arm length—for example, sliding the

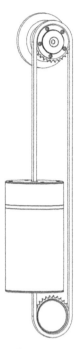

Fig. 6.2 An early design for rewind mechanism with the driving weight on an endless chain. © Charles Frodsham & Co

weight in or out—or by increasing the mass of the weight, which was considered an advantage in the initial stages of testing. Furthermore, this system does not use a suspended weight that could potentially swing at a detrimental frequency, producing what is sometimes referred to as the 'Thursday effect'. As a longcase clock is usually wound on a Sunday, by the time it is Thursday the weight(s) will have descended to approximately the height of the pendulum bob. If the clock case is not rigidly secured it (they) can start to oscillate in resonance to the pendulum, thereby robbing the pendulum of amplitude to the extent that it stops the clock altogether. Although the Burgess clock, with its large swing and its solid mounting to the wall, could never be stopped by the resonance of a relatively small driving weight, resonance of the weight could nevertheless badly influence the timekeeping.

There is one small disadvantage to the system, which is a change in driving force since the weight travels along part of an arc, but this

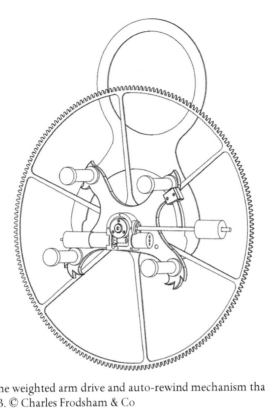

Fig. 6.3 The weighted arm drive and auto-rewind mechanism that was made for Clock B. © Charles Frodsham & Co

variation is less than half a per cent of the whole torque and occurs cyclically, besides which the remontoire removes much of this variation. It had been suggested that a weight hanging on a ribbon or chain from a circular cheek would remove this effect, but this again might have introduced parasitic oscillations.

The chosen system uses a shaft running in two shielded ball bearings in the great wheel bridge. On the back of the bridge, the aluminium housing rotates around the shaft, itself running on two ball bearings. Inside this housing is a worm wheel that is fixed to the shaft. A worm engages this wheel. It is fixed to the motor, which is attached to the case from the outside. On the housing, opposite the motor, are two mercury switches and a safety cut-off switch mounted in an enclosure. In front of the great wheel bridge, the hub of the great wheel is attached to the shaft using an expanding collet. The great wheel hub also allowed for the mounting of the double-ended hour hand, which was made to Burgess's design, having an aluminium moon and copper sun (Plate 21).

The mercury switches detect two levels of the weight arm. The lower one switches the motor on, and a higher one switches it off. The safety toggle switch comes into operation only if the switching off should not happen for whatever reason. It is operated by a pin on the back of the great wheel bridge.

To drive the rewind, an electronically commutated motor of 32 mm diameter with an attached planetary gearbox was chosen. The two-stage gearbox provides a reduction of 231:1, greatly increasing the torque available to drive the worm. Electronically commutated motors, also known as brushless motors, replace the brushes by electronics contained in an external control unit. This type of motor was chosen for its longer service life. Hall sensors in the motor gather the information of the angular position of the rotor and transmit it to the control unit. The unit itself has some sophistication built in. It is possible to set a certain speed of rotation and define a ramp for accelerating and decelerating. The unit also checks whether the motor is actually rotating and it keeps an eye on the current drawn, to pull the plug, so to speak, if an overload condition should occur.

One great feature of the chosen motor is its quiet running. The gearbox employs helical gear wheels made from a combination of Delrin plastic, steel, and aluminium, to reduce noise. To date, the system has run faultlessly.

With the various parts of Clock B received from Martin were the original holders for shaping the radii on the brass suspension cheeks on a lathe. As the Frodsham workshops are equipped with a high precision milling machine, it was decided to mill the radii, rather than turn them. Mervyn had suggested a radius of 3.5 inches on both cheeks. The two cheek blocks are mounted to a solid brass block, which in turn is attached to the steel backplate through two pillars (Plate 22). There was concern that a bimetallic effect between the steel plate and the brass could result in a distortion of the cheek block assembly and for that reason the attachment points were replanted closer together to reduce the effect.

The pendulum came with a suspension spring, which had been reduced in thickness by Martin. Martin and Mervyn thought temperature change should not alter the elasticity of the spring; they chose

Ni-Span-C to achieve that. Ni-Span-C is the trade name of an iron-nickel alloy introduced by the Special Metals Company, with small amounts of titanium, aluminium, silicon, and manganese added to give it its specific properties. Mervyn thought that to adjust the clock we should have suspension springs of various thicknesses to experiment with. Recent progress in technology has introduced a new generation of pressure sensors, which largely do away with mechanical components found in weighing scales and mechanical manometers. As a result, Ni-Span-C went out of fashion and for that reason we had some trouble finding a source. Eventually, with the help of Mervyn, we tracked down a supplier and acquired material in two different thicknesses. Cutting the strips to length could have easily set up stresses and distorted the ends of the spring, so to avoid that, we used our electrical discharge machine to cut the ribbon to length and also to create the two holes needed to attach the spring to the pendulum rod and spring block. The spring material was then submitted to a heat treatment in a protective atmosphere, to get the temperature coefficient of elasticity close to zero.

It is important to take note that the temperature coefficient of expansion of Ni-Span-C is not zero. Therefore, to avoid the suspension spring moving up and down in relation to the point at which the spring contacts the radius of the cheeks, the clock is supporting the spring with an Invar block mounted on a certain height of brass to give a combined expansion identical to that of the suspension spring. The expansion below the point of contact, as well as the expansion of the Invar pendulum rod, has to be compensated for by a brass sleeve underneath the pendulum bob. Martin and Mervyn made the temperature compensation beforehand, and we did not interfere with it as it proved to be incredibly effective.

To determine the rate of the clock, we integrated a Microset light sensor into the holder Martin had made. The holder arrangement fits nicely in a central gap of the beat plate. Finally, Peter Fox engraved the seconds and main dials, and the steel back plate—all 90 kilograms of it—was painted matt black. To keep dust and air turbulences out, a Perspex case was commissioned and mounted around the back plate onto a wall-mounted wooden frame. This case was not airtight but offered good protection against dust and draft. Eventually, the clock was put together and set going.

Test period

After finishing all the parts and sorting out a few minor problems with the remontoire pin cage, as well as some end-shake adjustments and adjustments to the driving weight, the clock was up and running.

In July 2011 data of Clock B's rate was started using a Microset timer, and on 11 October, the Microset was set to be controlled by a GPS time base. A further manual daily check on rate was also taken, using a Junghans radio-controlled clock. With some practice, it was possible to judge the time indicated by the clock to around ±⅕ second (Fig. 6.4).

Rating Certificate Continued

DATE	TIME OF WATCH PASSING H.	M.	S.	POSITION	DAILY RATE	REMARKS	Day since set on 8th Aug.
24.8.11	8	06	22		0.0	+⅘'	17
25.8.11	8	06	22		0.0	+⅘	18
26.8.11	8	10	22		0.0	+⅘'	19
30.8.11	9	35	22		0.0	+⅘'	23
31-8-11	14	47	21.5		−0.5	Probabley a progressive loss	24
1-9-11	9	50	21.5		0.0	+⅘' over a number of days	25
2-9-11	12	07	21.5		0.0	+⅘'	26
5-9-11	12	48	21.5		0.0	+¼'	27
6-9-11	09	40	21.5		0.0	+⅕'	28
7-9-11	10	13	21.5	−0.5	0.0	+⅕'	29
8-9-11	09	30	21.5		0.0	+⅕' overall loss of 0.5s in 31days	31
9-9-11	09	26	21.5		0.0	+⅘'	
12-9-11	09	15	21.5		0.0	+¼'	
13-9-11	09	10	21.5		0.0	+⅘'	
14-9-11	10	50	21.5		0.0	+⅕'	
15-9-11	09	12	21.5		0.0	+¼ perhaps nearer to 21.6 or 21.7	
16-9-11	09	36	21.5		0.0	+⅕'	
19-9-11	11	12	21.75		+0.25	+⅕' accumulated over about 7 days	42
20-9-11	09	24	21.75		0.0	+¼'	
21-9-11	09	08	21.75		0.0	+¼'	
22-9-11	10	02	21.75		0.0	+⅓'	
23-9-11	09	02	21.75		0.0	+¼' at 11-58 power cut just	46
26-9-11	09	15	21.75		0.0	+⅕' 2s or so,	
						Emergency power for computer	
						& winding plugged in 26-9-11	
27-9-11	09	15	21.75		0.0	+¼'	Day 50
28-9-11	09	00	21.75		0.0	+¼'	
29-9-11	09	05	21.75		0.0	+¼' Tending to lose - high temp 26-30°c	
30-9-11	08	49	21.75		0.0	+¼' then air con. dropping temp	
						rapidly	
3-10-11	08	40	21.6		−0.15	Pendulum shielded from	
4-10-11	08	55	21.6		0.0	direct sun light (& sensor shielded)	
5-10-11	09	01	21.6		0.0		
6-10-11	08	50	21.6		0.0		
7-10-11	08	30	21.5		−0.1		
10-10-11	09	26	21.5		0.0	+¼' Lost ¼sec over about 2 weeks	
11-10-11	09	08	21.5		0.0	+⅓'	Day 64
12-10-11	10	13	21.5		0.0	+⅓'	

Notes: Microset timer changed to GPS 11th Oct 2011

Fig. 6.4 The rating sheet for Clock B, recording daily rate observed between 24 August and 12 October 2011. © Charles Frodsham & Co

After some initial adjustments, the clock gained 2.5 seconds in the first fifteen days of its test run. It was adjusted by turning the fine-rating nut at the bottom of the pendulum rod by 1⅔ of a revolution on 28 July. The amount of adjustment was calculated using Martin's comment in the *Horological Journal*, which stated that one complete revolution of the fine-rating nut changes the rate by 1/10 second per day. The clock was set running again on 8 August 2011 and then ran continuously until 30 March 2012 (Fig. 6.5).

During this 211-day period, the temperature varied from an ambient room temperature to +30°C. In the first 100 days, the clock had an accumulated error of + 0.7 of a second. The maximum gain over this period was 0.7 seconds, and the maximum loss was 0.8 seconds. Allowing for reading errors, this was very encouraging. Over this period the Perspex cover was removed on two occasions for filming. Over the next 100 days the clock began to gain fairly steadily at around 1/10 second a day, resulting in an accumulated error of +8.4 seconds and a maximum daily error of +0.3 second. Interestingly, over the following 33 days, the rate settled again and the clock error was -1/10 second slow at the end of this period. The clock's ability to cope with temperature changes and disturbances, such as removing the cover, was impressive.

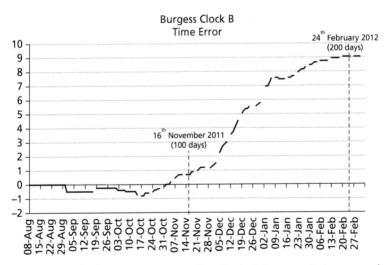

Fig. 6.5 Graph showing time error for Clock B between 8 August and 12 October 2011. © Charles Frodsham & Co

Clock B at Greenwich

The clock was delivered to Greenwich on 18 April 2012 for independent testing.

Reference

Burgess, M. (1996). The scandalous neglect of Harrison's regulator science. In: Andrewes, W., ed., *The Quest for Longitude: The Proceedings of the Longitude Symposium, Harvard University, Cambridge Massachusetts, November 4–6, 1993*. Harvard, Collection of Historical Scientific Instruments.

7

Adjusting and Testing Clock B at the Royal Observatory, Greenwich

Jonathan Betts

Introduction

This paper provides a narrative of the trials and adjustments of Burgess Clock B at the Royal Observatory, Greenwich from 2012 up until 2015, first briefly contrasting this author's experiences in rating and adjusting conventional regulator technology. The paper explains in simple, practical terms how Harrison's whole precision pendulum system works and comments on the extraordinarily successful results of the trials.

Background

The story of the commissioning and creation of the two Harrisonian Burgess clocks, A and B, in the mid-1970s, and the purchase of Clock B by Don Saff and its completion by Charles Frodsham & Co. in 2010, are covered elsewhere in this volume. Following the reports of Clock B's fine performance at the workshops of Frodsham & Co., horology staff at the Royal Observatory, Greenwich were naturally very interested when invited to carry out independent timekeeping trials of the clock at the Royal Observatory in Greenwich in 2012.

My own connection with the clock goes back to its origins in the 1970s when, as a self-employed clockmaker recently set up in business in Ipswich, I was commissioned by Martin Burgess to make the brass bearing boxes for both clocks, housing the ball races in the train and remontoire. It is important to state that at that time, while I was a great admirer of Harrison's achievements, particularly of his pioneering work on chronometry, and of Martin Burgess's creative work, I was frankly

Betts, J., *Adjusting and Testing Clock B at the Royal Observatory, Greenwich* In: *Harrison Decoded: Towards a Perfect Pendulum Clock*. Edited by Rory McEvoy and Jonathan Betts, Oxford University Press (2020).
 DOI: 10.1093/oso/9780198816812.003.0007

sceptical about Harrison's claims for his pendulum clock technology. Only a few years before I had undergone traditional British Horological Institute (BHI) training in technical horology at Hackney College, and everything we were taught concerning optimum precision pendulum clock design indicated that the Harrison system could not perform particularly well.

The grasshopper was a recoil escapement, and was controlled by a relatively light pendulum that ran at the very large arc of 12°. Worst of all, the pendulum suspension ran between suspension cheeks. With a precision pendulum, the one thing one should not do, we were taught, was to mess about with the pendulum's *length*, and this is what suspension cheeks do constantly when the clock is running. Besides, cheeks had been tried and proved useless years before Harrison's time and, we were told, Harrison was on a hiding to nothing with his pendulum clock theories. Such clear, unequivocal instruction, given in a formal educational environment, is not readily forgotten or questioned. Certainly, Harrison's remark that his Late Regulator should be capable of keeping time to within one second in 100 days was simply dismissed as fantasy by most commentators, including some of the most respectable and admired of horologists at the time.

Having said this, the subsequent thirty-five years of work on precision pendulum clocks at the observatory had introduced doubts in my mind about received wisdom, too. To my growing disappointment over the years, I found that even the very best regulators with deadbeat escapement (and with every other refinement prescribed in established thinking) did not produce nearly as good and stable long-term time-keeping as expected. I began to doubt the accepted belief that a good regulator is routinely capable of keeping time to within ±1 second a week, long term.

Given that an ordinary longcase clock with a seconds pendulum and anchor escapement can, if cared for and kept in a reasonably constant temperature, easily keep time to within half a minute a week, long term, I began to wonder whether the better timekeeping of a good regulator was actually owing to its quality of construction, its compensation pendulum, and its solid and stable mounting. Was the deadbeat escapement itself actually an improvement at all? John Harrison certainly didn't think so, famously remarking when he first saw George Graham's escapement that he thought 'either he must be out of his Senses, or I must be so!' (Harrison, 1775:7).

A long-term experiment was then conducted on a fine regulator with jewelled deadbeat escapement at the observatory, by alternately running the clock with its deadbeat escapement and then with an equivalent anchor recoil escapement, all else remaining the same. The anchor pallets were identical to the deadbeat ones except they were in hardened steel (perhaps thus giving the anchor escapement a slight disadvantage) and had the pallet faces designed to provide about one degree of recoil on the wheel during the usual supplementary arc. The trials of this clock continued for over four years and showed very little difference in average timekeeping stability between the two escapements, supporting my view that, at least over the period of the experiment, there was no significant technical advantage in the deadbeat clock escapement. [Editor's note: the author wishes to inform readers that this research will be published in the near future.]

As told elsewhere in this volume, from the late 1970s a group of horologists, known informally as the Harrison Research Group, formed with the intention of looking at Harrison's work and ideas. The views of members of the group, and some notes on the performance of Clock A before its delivery to Norwich, were well known by this time, and thus when Burgess Clock B came to the observatory for independent time-keeping trials in 2012, it came with certain tacit expectations. The principal focus for scrutiny was Harrison's (1775:35–36) remark in that his pendulum clock system should be capable of keeping time to within one second in 100 days.

Harrison's remark was stark in its boldness. Keeping time to within a second in 100 days is quite clear. No rate is to be applied to this clock, and the error, Harrison predicts, simply will not exceed one second in either direction over a 100-day period. It was a bold claim in the eighteenth century—after all, even today the majority of precision pendulum clocks, for which long-term stable timekeeping is claimed, have rates applied to the data. So, if Harrison's prediction was proved right, it would mean that such a clock would be directly comparable with the best pendulum clocks produced in the twentieth century, including the celebrated designs by Riefler, Shortt, and Fedchenko.

While Clock B is not in any sense a replica of Harrison's Late Regulator, the scepticism expressed by the horological establishment principally concerned the *general principles* of Harrison's pendulum clock design. With those general principles incorporated into Clock B, comparisons with clocks that are protected from changes in air pressure, such as the

Riefler and Shortt systems, were thus waiting to be made as the clock revealed what it was capable of.

Installation and the location of the clock

In 2012, the clock was set up at the observatory, with the assistance of the Charles Frodsham & Co. team. The site chosen was the horology workshop in the Great Equatorial Building—a room viewable to the visiting public through a glass-backed showcase running across the full width of the room (Plate 23). This enabled the clock experiment to be seen by visitors whenever the museum was open, whilst ensuring high security (access inside the workshop itself was limited to National Maritime Museum (NMM) staff and their visitors). The clock was fixed securely to the south pillar support of the Great Equatorial Telescope, a solid stone and brick support buried into the ground below and the most stable position in that area. Nevertheless, it must be noted that this position was not without its environmental issues.

On the question of physical stability, while the clock was attached to the solid brick and stone column, in the years since the 1850s when the column was first built, a large mature beech tree has grown up just fifteen feet away on the south side of the building, and the roots of the tree must surely reach the base of the column. A few feet further away, on the other side of the observatory's boundary, are several mature chestnut trees that must also create small disturbances to the support in windy weather. During the whole period of the trials of Clock B, the workshop continued to be in regular use for horological maintenance. From time to time the workshop was the venue for horological tours as well as regular press and TV interviews, and no attempt was made to isolate the clock from the everyday activities around it. The Great Equatorial Telescope itself was occasionally in use and any vibrations caused by the motion of its several tons would have been transmitted directly down the column to the clock. In addition, having the museum's visitors passing just a metre or so from the clock every day, with the double doors into the gallery sometimes banged open and closed, will also have created small disturbances in the support.

Regarding temperature, the workshop had to remain a habitable space for the horology staff and was heated in wintertime. In the summer, if the room became too hot, a ventilation fan was switched on to circulate the air. However, the room was not in any other sense

Plate 1 William Hardy's transit clock, now much-altered and mounted within a stone pier in the Airy Transit Circle at the Royal Observatory, Greenwich. Credit: NMM.

Plate 2 The Shortt free pendulum system (No.16) made around 1926 by the Synchronome Company Ltd. Credit: NMM.

Plate 3 The Harrison oak-framed clock at Brocklesby that does not require lubrication. Image courtesy of Dr John C. Taylor.

Plate 4 Harrison's manuscript drawing of an escapement that is likely to have been the one he made for Trinity College, Cambridge in 1755. Credit: NMM.

Plate 5 Pallet frame for the Brocklesby clock, showing the remainder of a mortice similar to those depicted in the manuscript drawing. Image courtesy of Dr John C. Taylor.

Plate 6 Harrison's 'double-thrust' version of the grasshopper escapement (Brocklesby clock). Image courtesy of Dr John C. Taylor.

Plate 7 Burgess Clock B's co-axial grasshopper escapement. Credit: NMM, by permission of Don Saff.

Plate 8 A later gridiron pendulum from a clock by John Shelton, London that was made to time observations of the 1761 transit of Venus (please note that the iron piece was added in the twentieth century to enable electromagnetic correction). Credit: NMM.

Plate 9 In this photograph of Mulberries, taken around 2009, the original house is on the left; the old chicken coop, in which Martin constructed his first sculptural clock (the Broxbourne clock), is on the far right; and the new workshop, built around 1965, is in the middle. Photograph © William Andrewes.

Plate 10 Martin in his workshop, working on the great wheel of the Schroder clock, *c.*1968. Photograph © Martin Burgess.

Plate 11 Martin and Eleanor Burgess on 2 August 2001. In addition to clock-making, a large amount of their time was devoted to tending the land, off which they lived for about fifty years. Photograph © William Andrewes.

Plate 12 Martin Burgess's second commission clock, known as 'Simon's Balls'. © Don Saff.

Plate 13 The clock was constructed in 1969 for the merchant banking firm Schroder Wagg's new premises in Cheapside, London. © Don Saff.

Plate 14 The Gurney clock in its last display (removed for storage in 2015) at the Castle Mall in Norwich, Norfolk. Photo by Leo Reynolds.

Plate 15 Martin Burgess and Don Saff at Mulberries in 2009. © William Andrewes.

Plate 16 Meeting at Punnets Town, 28 June 2011. From left to right: Martin Dorsch, David Thompson, Daniela Hofer, Richard Stenning, Roger Stevenson, Jonathan Betts, Andrew King, Will Andrewes, Philip Whyte, Don Saff, and Ruth Saff. © William Andrewes.

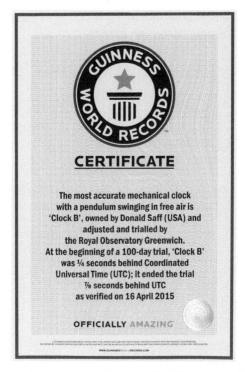

Plate 17 The certificate issued by Guinness World Records in 2015, following the peer-reviewed 100-day trial earlier in the same year. Credit NMM.

Plate 18 Clock B at the workshops of Charles Frodsham & Co. prior to commencement of work. © Charles Frodsham & Co.

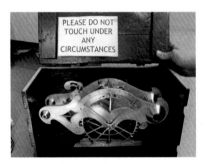

Plate 19 Clock B's escapement module, in its protective box with a severe warning to the unfamiliar from Burgess. © Charles Frodsham & Co.

Plate 20 Components of the escapement and remontoire, prior to modification. © Charles Frodsham & Co.

Plate 21 Fitting the Sun and Moon hour hand to the great wheel hub.
© Charles Frodsham & Co.

Plate 22 Showing Clock B's original brass cheeks and pendulum suspension.
© Charles Frodsham & Co.

Plate 23 Clock B in the Horology workshop at the Royal Observatory,
Greenwich.

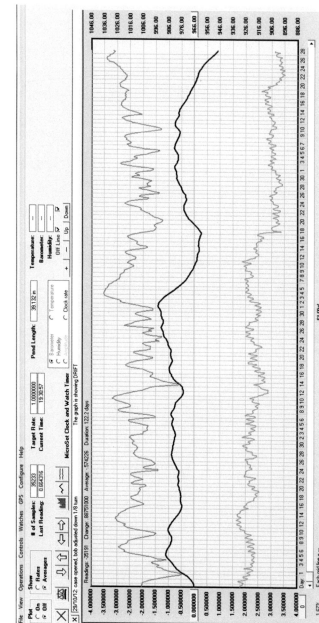

Plate 24 The Microset readings from the beginning of the first trial, from 29 October 2012, through until the end of February 2013. The clock error (dark blue line) stayed within one second for most of that four-month period, though exceeded it after the first week and during the middle of that period. Owing to the way the Microset compares with the time standard, the 'minus' seconds figures for error on the left-hand vertical scale indicate a positive error shown on the clock. Credit: R. McEvoy.

Plate 25 Jonathan Betts with Sir George White, Keeper of the Clockmakers' Company Museum, sealing the clock case. Inset: Sir George White's seal on the clock.

Plate 26 Peter Wibberley of the National Physical Laboratory (NPL) sealing the clock case. Inset: The NPL seal on the case.

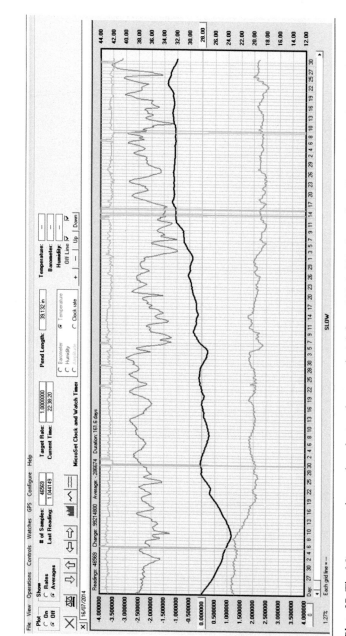

Plate 27 The Microset readings taken during the second trial, from 23 July 2014 through to 1 January 2015, showing clock error (dark blue line). Please note that in generating this graph from the Microset data, the indication for clock error beginning at 23 July originates on the graph at '0' seconds, whereas at this point in the clock's trial (nearly four months after the trial began) the clock was in fact displaying an error of almost exactly minus one second on UTC at that point. Therefore, if the seconds scale is referred to on the left of the graph, one second needs to be subtracted from the apparent error. Credit: R. McEvoy.

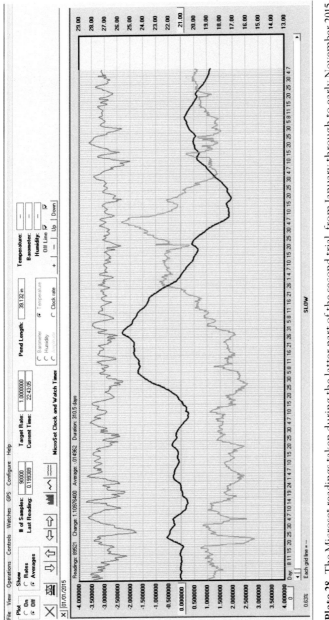

Plate 28 The Microset readings taken during the latter part of the second trial, from January through to early November 2015. At the beginning of this period, nine months after the display case was secured shut, the clock error (dark blue line) was back reading zero, and by early November was back reading just over a second slow. Credit: R. McEvoy.

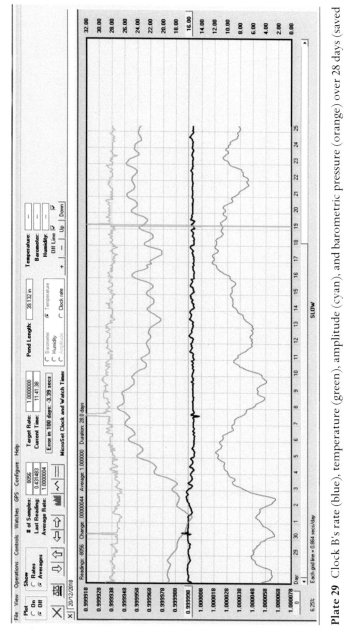

Plate 29 Clock B's rate (blue), temperature (green), amplitude (cyan), and barometric pressure (orange) over 28 days (saved 25 May 2018). Credit: R. McEvoy.

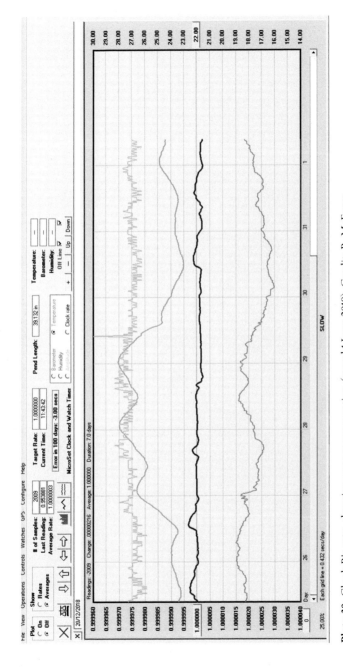

Plate 30 Clock B's rate, showing overcompensation (saved 1 June 2018). Credit: R. McEvoy.

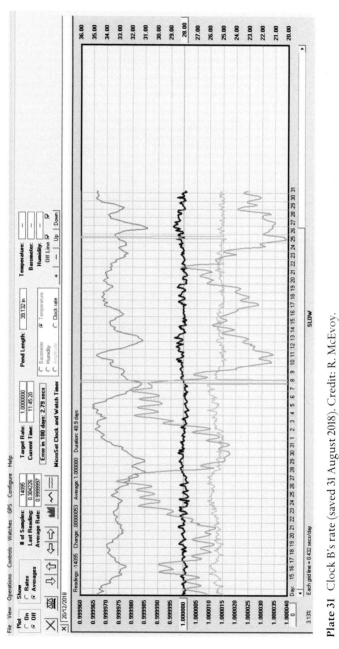

Plate 31 Clock B's rate (saved 31 August 2018). Credit: R. McEvoy.

temperature controlled. There was no air conditioning available, and over the course of the trials the temperature ranged over more than 10°C.

The clock was given physical protection by a Perspex case enclosing it against the wall, but this was not hermetically sealed and an opening at the base of the case (providing access for cables feeding the monitoring equipment) enabled free flow of air between the clock and the workshop surroundings.

The trials

The first trials, which began immediately after installation through to 2015, were conducted by myself, with the assistance of colleague Rory McEvoy, who was entirely responsible for maintaining and adjusting the monitoring equipment. This was Bryan Mumford's Microset device, which was set up to record twenty-four hours a day, seven days a week, with the data stored on a dedicated PC alongside, on an uninterruptible power supply. The Microset employed an infrared sensor that read the passing of the pointer at the bottom of the pendulum, recording the error/rate of the clock, as well as a nominal amplitude for the pendulum. Additional sensors inside the Perspex case recorded the temperature, the barometric pressure, and the relative humidity.

With this equipment in place the clock was quickly brought to time, and as the staff at Frodsham's had found, it was soon clear that the clock had a very stable rate. Martin Burgess was naturally very interested in following its progress, and reminded us that although the clock may appear to be going well, it had not been adjusted for barometric pressure (air density) changes, so we should not expect too much of it. (Harrison's adjustments to compensate for changing air density will be explained later in this paper.)

After just a few weeks' running the clock, it was evident that while the temperature compensation appeared good, the rate was indeed affected by barometric pressure changes, as Martin had predicted.

Notwithstanding the need for Harrison's air density compensation, the rate of the clock appeared so stable in other respects, the decision was taken to close the case and give the clock a 100-day trial. The clock had been rated to zero at an average barometric pressure (around 1010 mb was chosen) and temperature, and it was hoped that the effects of rising and falling air density would 'give and take' sufficiently to keep the clock error near zero.

The first trial, 2012–2013

In the following days, however, in late October, a steep fall in the barometer caused the clock to gain and within a week of the test beginning, the error shown on the seconds dial had exceeded one second. As the Microset graph from this period shows (Plate 24), with the fall in barometer, the clock's error went beyond one second early on in the trial. Following that point the clock continued to respond to pressure changes by losing and gaining as the barometric pressure went up and down, but it did manage to give and take as had been anticipated. Most importantly, there were no signs of unpredictable instabilities or significant drift of rate and, after the first few days of the trial, the error of the clock remained within one second of mean time for 110 days. The results looked promising—the clock had already achieved a period of remaining within one second in 100 days—and it was clear that if air density effects caused by barometer change could be removed, the performance of the clock should be very remarkable indeed.

Harrison's 'barometric' (air density) compensation

The time had come to adjust Clock B to introduce Harrison's air density compensation, which meant carrying out what has been termed 'hill tests'. In recent years, a number of interested horologists have remarked that these tests appear to them to be very complicated—this is not so, and the principle is in fact very simple to understand. In explaining the way the compensation works it is important to understand first what happens to a pendulum when the air density changes. Such changes are caused by barometric pressure change but also, entirely separately, by a change in air temperature.

If the temperature of the air falls, the air density will increase, causing the pendulum to lose. However, this temperature effect is the *opposite* of the temperature effect on the length of the pendulum and the air density effect somewhat reduces the larger effect and reduces the required temperature compensation in the pendulum. In addition, the effects on air density of changes in barometric pressure and temperature can, to some extent, cancel each other out or can compound.

How, then, does changing air density affect a swinging pendulum? If the air density increases, two things happen: first and foremost, with a rise in air density, the pendulum experiences increased floatation effect. Just as with a body suspended in water, it is buoyed up more and this effectively reduces the value of the restoring force gravity, so the pendulum tends to swing more slowly. Secondly, owing to the greater density of the air, the pendulum has more work to do in passing through it, and the greater energy consumed means that the arc of swing reduces. The amount it reduces depends on the shape and size of the pendulum, and how quickly it changes will also depend on the sheer mass of the pendulum. A very aerodynamic and very heavy pendulum will not change its arc so much, nor so quickly, with a change in air density.

Because of circular deviation, minimising arc change might seem like a good thing and traditional theory proposes maximising the aerodynamics to achieve this. However, whatever the shape of the pendulum, it will be subject to the first problem of floatation effect, which without compensation will always cause the pendulum to swing more slowly when the pressure rises. This is why traditionalists then fit some form of compensation, like an aneroid capsule, to correct for floatation.

Harrison looked at the problem in an entirely different and holistic way. He accepted that the floatation effect is present, but recognised— indeed, welcomed—the fact that the arc will also change owing to increased air density. Harrison's pendulum system incorporates suspension cheeks and, with circular deviation removed by them, he then deliberately reintroduced a measured amount of circular deviation, specifically to counteract the floatation effect. For this reason, the Harrison system actually avoids a highly aerodynamic pendulum, because it needs a significant change in arc with air density change in order to get the correction from the suspension cheeks. As explained elsewhere in this volume, the presence of Harrison's recoil grasshopper escapement also plays its part in the dynamics of this whole process, but the above provides an explanation of the rudiments of the compensation system without going into too much detail here.

A simple explanation

In order to introduce this air density compensation, it is necessary to adjust the suspension and cheeks, so that, all other things being equal, there is a slight *gaining* tendency when the arc falls. This is where the hill

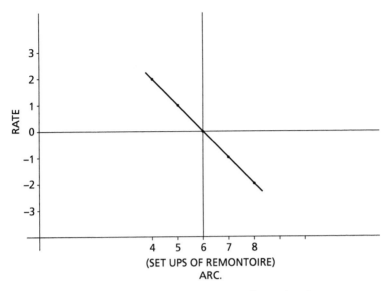

Fig. 7.1 A simple graph showing arc against rate. © Jonathan Betts

test comes in—the term 'hill' referring to the shape of the curve on the graph resulting from the test. What the hill test does is simply to record the rate of the pendulum at a variety of different arcs, on either side of the running arc.

Figure 7.1, which shows rate (vertical axis) against arc (horizontal axis), is not a real recording of any data; it just shows in simple terms the kind of adjustment being sought. In reality, this graph is one side (the right-hand side) of the 'hill'. In order to carry out the test, and create such a graph, one has to artificially change the arc of the pendulum in a very repeatable way, and to do this on a Harrison-type clock, one changes the driving force to the escapement by changing the initial set-ups of the clock's remontoire.

Figure 7.1 shows readings at five different arcs, which is more than sufficient for the test. For the actual hill tests the clock was tested with a greater number of different arcs. With the remontoire at four setups, the clock would barely run at its escaping arc of about 10.5°. At ten set-ups of the remontoire, the pendulum was running at the limits of the pendulum scale, at an arc of over 15°.

The test at various arcs has to be done quite quickly, to ensure that they are all carried out at the same barometric pressure and temperature. If the pressure or temperature changes significantly during the tests, then the results will be flawed. What is shown in the graph (Fig. 7.1) is an increasing gain at lower arcs of the pendulum on lower remontoire setups. In this way, when the pendulum arc falls owing to a rise in air density, the losing tendency caused by the floatation effect is cancelled out by the gaining effect in the cheeks.

Naturally, the system depends on the arc of the pendulum *only* changing when the air density changes. If the pendulum arc changes for any other reason, such as a change in impulse, then there will be a small unwanted change in rate. This is why Harrison employed the remontoire, to ensure that the impulse was as constant as possible.

The practice of the hill test

In arranging for the clock to gain in the smaller arcs, the pendulum and suspension are adjusted in two ways. First and foremost, the suspension cheeks have to be cut to the optimum radius to put the adjustments in the right ballpark, with the system close to, but not quite, isochronous, with the pendulum running slightly fast in the larger arcs. For Clock B this had already been done during completion at Frodsham & Co. The second, somewhat more subtle means of adjustment is then to alter the thickness of the suspension spring. As Harrison instructed, it must be 'thin to the purpose'. Changing the thickness of the suspension spring has a slightly counterintuitive effect, as a thicker spring produces a greater loss in the larger amplitudes. One might expect a thicker suspension spring to introduce a greater net gain into the system, but experiment shows this not to be the case. This would seem to be because a thicker spring wraps less closely to the cheeks and less correction is applied. In carrying out the hill tests on Clock B, four different thicknesses of Ni-Span-C suspension spring were tried.

The thinnest spring fitted was five hundredths of a millimetre. In fact, this spring was so thin that the path of the pendulum was quite unstable and it was unlikely to be a viable suspension, but it was possible to derive a trace for the hill test. The resulting graph (Fig. 7.2) showed a steep incline in the wrong direction—the higher the amplitude, the greater the gain.

A thicker spring of nine hundredths of a millimetre was tried, and this resulted in the beginnings of a hill on the graph (Fig. 7.3), but the position of the running arc, at six setups of the remontoire, was still on the 'ascent' of the hill, which was not what was needed.

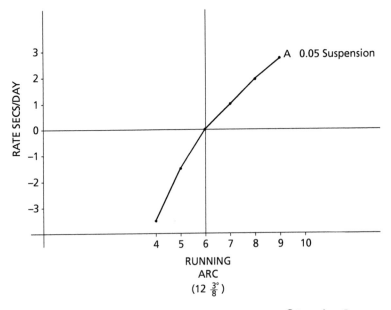

Fig. 7.2 The Hill Test graph for the 0.05-mm suspension. © Jonathan Betts

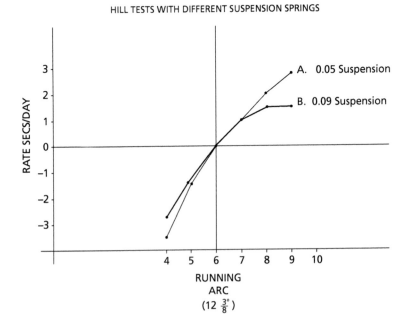

Fig. 7.3 The Hill Test graph for the 0.09-mm suspension. © Jonathan Betts

HILL TESTS WITH DIFFERENT SUSPENSION SPRINGS

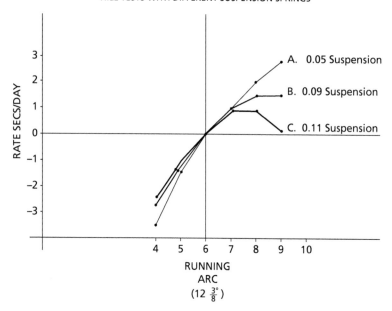

Fig. 7.4 The Hill Test graph for the 0.11-mm suspension. © Jonathan Betts

With a spring of eleven hundredths of a millimetre (actually the spring which was supplied with the clock), one begins to see (Fig. 7.4) the hill properly forming, but it is still too far over to the right of the running arc. It was clear a thicker spring was needed and Frodsham & Co. were asked to source some new Ni-Span material somewhat thicker than what we had been using.

In fact, the thickness of what they provided was not much greater, at just twelve and a half hundredths of a millimetre, but the Ni-Span type alloy, although nominally of the same type, appeared to have a distinctly greater modulus of elasticity—it was noticeably stiffer, probably owing to a different treatment during manufacture.

This provided what we needed, and the result of that hill test (Fig. 7.5) showed a gaining tendency as the arc reduced below the running arc and a loss when the arc increased. In theory then, this should provide some correction for changing air density. (This perfectly logical series of actions, leading to a successful conclusion, has curiously been described by some commentators as our having stumbled on the

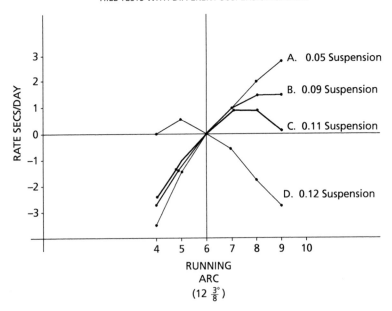

HILL TESTS WITH DIFFERENT SUSPENSION SPRINGS

Fig. 7.5 The Hill Test graph for the 0.12-mm suspension. © Jonathan Betts

correct adjustment 'by chance'. It is difficult to imagine how, or why, such a conclusion could reasonably have been drawn.) With the new thicker twelve-hundredths suspension spring in place, the clock was duly brought to time, the Perspex case was closed and a new run was started for the clock to see how it responded.

The second trial, 2014–2015

As before, we were immediately struck by the apparent stability of the clock's rate, but this time things progressed rather differently. Soon after the test began there was a significant fall in the barometer, just as we had had on the previous test, but this time the clock's rate appeared largely unaffected. This exceptionally stable rate was very encouraging and surprising, and we soon resorted to manual double-checking with the telephone speaking clock, just to be sure we were not deceived.

After about ten days of seeing no perceptible error appear at all, in spite of significant change in the barometer, the decision was made to go for an unofficial 100-day test. If the clock continued to perform as well as it had done in the previous few days, its longer-term stability would be almost incredible, so the decision was taken to have the case independently locked shut with wax seals on the cross-drilled fixing bolts. In this way, should the clock perform as well as hoped, there could be no suggestion that anyone had tampered with it.

Two different organisations were asked to seal the case, the first being the Clockmakers' Company, in the form of the Keeper of its Collection, Sir George White, who sealed one of the bolts with his own family seal and signed a certificate to that effect (Plate 25).

Later the same day, the NMM Director attended and Peter Wibberley, Senior Research Scientist of the National Physical Laboratory, came to seal the case with the NPL's official seal[1] (Plate 26). One hundred days later saw the first of the Greenwich symposia and, as revealed to the delegates that day, the clock was still showing an error well under one second from UTC. Following the symposium, the clock remained closed with its wax seals and, apart from one or two occasional glitches in the monitoring equipment, the data continued to be collected through most of 2015.

Guinness World Record

The period including the autumn and early winter of 2014/15 saw a separate independent observation of the clock's timekeeping, which continued to show variations less than one second in 100 days. This short trial within the longer trial then qualified the clock for a Guinness World Record of 'Most Accurate Mechanical Clock with a Pendulum Swinging in Free Air'.

The clock was allowed to continue running through 2015, past the second Greenwich conference, and into the following year. During that time, there were several separate periods of 100 days when the clock stayed within one second of UTC.

---

[1] That day was the only one on which both witnesses could attend, but it is unfortunate that it just happened to be the 1st of April. It must be pointed out that there is no significance in this coincidence and that everyone concerned took the proceedings very seriously indeed!

However, it must be pointed out that over that period the clock's rate was not entirely static and the part of the Microset graph from this latter part of the second trial (Plate 27) shows that at one point the clock didn't stay within a second of UTC. By August 2014 the clock was reading -2 seconds but then, with significantly cooler weather, the error of the clock turned back. By mid-November, it was reading close to zero again and its error continued to be close to zero, showing about half a second out on 18 April 2015 (Plate 28), the day of the second Greenwich conference. Closely following the clock's behaviour, it was now clear that, following the adjustments for air density compensation, the clock now had a small temperature coefficient.

Having already established in the first trial that the temperature compensation in the clock's pendulum was all but perfect, the reappearance of a temperature-sensitivity requires explanation.

What is seen here is, in fact, a direct result of introducing the air density compensation previously described.

Shorter when warm

It is important to note that, with a change in temperature, the effect on the clock's rate of the changing air density (and consequent change in floatation effect) is the *opposite* of that of a change in length of the pendulum rod.

With a rise in temperature, for example, an ordinary pendulum gets longer and runs slower, but at the same time the air density reduces and floatation decreases, effectively increasing the effect of gravity, causing a slight gaining tendency. This is proportionally much smaller than the effects of the linear change in the pendulum rod, but it is there and cancels out a small part of the need for compensation. When air density changes with temperature, the pendulum arc will also change, but with Harrison's suspension cheeks ensuring the pendulum is close to isochronous, this makes little difference in the first part of the testing procedure.

However, if one then introduces a more general means of compensating for the effects of changing air density by introducing a little circular deviation into the suspension, this not only removes the effects of air density change with barometer, but also temperature, and that cancelling out of some of the effects of temperature on the pendulum rod is no longer present. The result is that one sees a small temperature coefficient reappear in the system.

Air viscosity

In fact, the reappearing temperature coefficient would be somewhat greater still if it were not for the effects of air viscosity change at the same time. Air viscosity, quite separate from air density, causes frictional drag on the pendulum as it moves, and thus causes a small additional energy loss in the system. However, the viscosity of the air changes in a somewhat counterintuitive direction: In higher temperatures the air is more viscous, and consequently the net energy lost in a pendulum swinging through warmer air is greater. This means that the pendulum's arc in heat will be somewhat less, and the suspension cheeks will provide a small additional gaining correction in heat thanks to the changing air viscosity, aiding the compensation. Nevertheless, this effect is very small and is insufficient to cancel out the air density effects of temperature change.

Therefore, the compensation Harrison introduced for air density correction potentially leaves the system temperature-sensitive again. However, this temperature coefficient appears to be reasonably linear in its effect and can thus be largely compensated for by adding extra temperature compensation into the pendulum. This was all understood by Harrison, who stated unequivocally that the pendulum 'should be shorter when warm', though he recognised there might be a sensitivity in this area, stating clearly that, notwithstanding the clock's temperature compensation, it should ideally be situated in 'a pretty temperate place' (Harrison, 1763:41).

The end of the trial

Following the second Greenwich conference in April 2015, the clock continued to run for another year, and in spite of its lack of complete temperature compensation, continued to perform excellently, its maximum deviation from UTC during the whole two-year trial being just over two seconds at the end of May 2015. On 1 April 2016, precisely two years after the clock was set precisely to UTC and the case sealed with the wax impressions, the clock was inspected by eight members of the

NMM staff, including NMM Director Kevin Fewster, and all signed a certificate to record what they found. The group confirmed that the wax seals on the clock's case had not been tampered with and agreed on inspection that the clock was reading one second in advance of UTC. Allowing for the fact that a leap second had been introduced to UTC at the beginning of the year, this meant that the clock was actually showing no error at all.

The seals were then broken and the clock was subsequently moved to a new position in the Longitude Gallery at the observatory, an even more challenging location for a precision pendulum clock. At the present time (2018) the clock, while not on formal trial, still performs well and has since had an adjustable temperature compensation fitted to the pendulum, hopefully enabling further improvements to its timekeeping. The next chapter in this extraordinary clock's story begins.

Appendix: Critical comments

From their first creation in the mid-1970s, Burgess Clock A and Clock B attracted comment and interest owing to their wholly different approach to precision pendulum design. Owing to these radical differences, critical opinions have been expressed ever since, and predictions made about the probable capabilities of such a design. Comparisons were naturally made with traditional high-precision pendulum clocks such as those by Riefler, Shortt, and Fedchenko.

There have also been critical comments made in recent years in the horological press, specifically relating to the trials of Clock B. As these comments were current during the three Harrison symposia, for which this volume represents the proceedings, and as some were, in fact, voiced during questions at the symposia, it is appropriate to address these comments here.

It is also important to note here that while the authors of the chapters in this volume all share a common wish to encourage a better understanding of Harrison's horological philosophy and its superiority, the opinions expressed in the chapters are those of their respective authors and not necessarily that of the other contributors. It will be evident from their various backgrounds that the authors have studied this subject from a variety of different perspectives, each of which hopefully provides a different and interesting viewpoint.

Clock B

The trials of Clock B have shown that, in terms of long-term timekeeping stability, the basic philosophy of Harrison's precision pendulum system, as fitted to Clock B, has the potential to equal the performance of the twentieth-century creations mentioned.

But, in making such comparisons, it's also important to compare like with like. The timekeeping performance of clocks such as those twentieth-century examples mentioned were derived from trials conducted with those clocks in the best possible, physically stable, temperature-controlled environment. On the other hand, the trial of Clock B, as described earlier in this volume, was undertaken in a busy, non-air-conditioned workshop, with a public gallery and outside doors a few feet away and the mounting column close to large trees just outside the building.

Mathematical analysis

It's also important to remember when making comparisons with these 'high performance' twentieth-century clocks, that it is highly likely, given the national and commercial sources for some of those comparative data, that the best results were selected, and it must be pointed out that none of the test data routinely cited in the horological press come from *independently monitored* trials such as Clock B's. It is also highly likely that the figures quoted for those twentieth-century clocks have been mathematically manipulated. For example, rates are routinely determined and applied during scientific analysis of such precision clock results, whereas there was never any need to consider applying a rate to the Clock B data.

Mathematics can be a very seductive assistant, but there is the ever-present danger that it can lead one from the ideal, which is surely to design a clock which simply maintains correct time. There have to be limits in using maths to 'interpret' a clock's variable performance. After all, the time shown on the dial of every mechanical clock merely represents that clock's perfect response to the natural influences upon it, and what we call 'errors' in what the clock says are merely the result of influences we either haven't compensated for in the clock's design, or allowed for in interpreting the clock's data. If we go down the ultimate mathematical route, every clock will be a perfect timekeeper—entirely missing the point.

Clock B: what can be claimed?

Ultimately, has the trial of this clock proved that Harrison's Late Regulator would have performed to a second in 100 days? Of course not, no one can prove that question either way now, as that clock is over 270 years old and anyway is incomplete.

Does the design of the Late Regulator have the potential to perform to one second in 100 days? We cannot yet say, but we should be able to make an informed judgement on that question once exact replicas are made of Harrison's regulator, and we await with interest the completion and testing of a series of accurate replicas currently being made by members of the British Horological Institute.

Does Harrison's overall philosophy for the optimum design of a precision pendulum clock—with the recoil grasshopper escapement impulsing a relatively low-mass pendulum swinging through a large arc, as fitted to Clock B—have the potential to keep time to within one second in 100 days? One can definitely answer yes to this question; it does not require more than one clock, nor more than one trial (especially of over three years), to be able to say that the system has such potential.

There are, however, those who have expressed doubts about the performance of the Clock B itself. The decision to have the case sealed by independent bodies of unimpeachable impartiality at the beginning of the trial proved essential in this respect. Nevertheless, it has been suggested that, in fact, the results we have seen over those several years are more likely to be the product of chance or serendipity, the multitudinous and labyrinthine influences on such a clock being too complex to compensate for to such a degree of success. Proof positive, it is said, can *only* come from the application of a scientific method, where several such trials are conducted successfully with several different clocks.

This argument has also been applied to the trials of H4 when qualifying for the Longitude Award, and the same answer can be given. Strict scientific method is entirely appropriate when a single pure scientific concept with a single result requires verification. It is essential in fundamental studies such as in determining a chemical reaction, or biological testing. But in complex technological trials, where the results may have any number of different values and valid results, it becomes much less appropriate. How many sets of 100-day trials must a clock undergo successfully before one can agree that the design (not just that one clock) has the potential to succeed? Are such critics really suggesting that the one clock might be an extraordinary fluke and that in general others made in precisely the same way would not work well?

These commentators seem to be unwittingly echoing the words of the notoriously critical Thomas Earnshaw (1749–1829), who said of other chronometer makers that if their chronometers went well it is only because, unknown to them, one error must be compensating for another! Really, over the duration of these trials, the probability of such a thing happening is very small.

Was it needed?

If, as now seems highly likely, Harrison's regulator design is able to perform significantly better than a typical deadbeat regulator of the eighteenth century, it would be true to say that if astronomers such as Nevil Maskelyne had listened to Harrison, and had encouraged him to complete his regulator and construct others, eighteenth-century science would have had a vastly improved time standard.

However, it has been stated by one commentator that even if the Harrison system had been a significant improvement, the eighteenth-century scientific community actually didn't *need* or *want* a better time standard and were content with what they had. Such an extraordinary claim must be remarked upon and the myth dispelled (Bateman, 2017:26).

As told in Rory McEvoy's chapter, the scientific community in the eighteenth century, especially in France and England, certainly wanted better timekeepers—we know that throughout the history of science, practitioners like Maskelyne were always looking for better, more accurate instruments. Specifically, in Maskelyne's example, he and the Board of Longitude on which he sat were constantly searching for improved designs for pendulum regulators, and were prepared to spend considerable sums of money to encourage better designs. This was a practice which continued throughout the nineteenth century with George Airy and into the twentieth century.

Every observation on the transit instrument required not only accurate measurement of declination, but of right ascension—the sidereal time at which the body being observed transited the meridian. The more accurate the regulator, the more accurate the observation; errors in the regulator translated directly into errors of star position, so accurate timekeeping was vital. It is true that errors in the transit clock can be corrected with regular observations of 'clock stars', but cloudy skies often prevented this and, more importantly, clock correction was time consuming and every time saw the risk of human error in observation. The fewer times this was necessary the better, and a transit clock with improved long-term stability was constantly in demand.

Impossibly complex and expensive?

Another accusation which has been levelled at the Harrison system of precision pendulum clock design is that the Harrison-type clock would have been impossibly expensive and impossibly difficult to set up and adjust. This mistaken idea may have originated in the problems which we know Harrison had in developing the system, and which Martin Burgess experienced in his re-creation of the principles in his clock. However, now that the design is finished and tested, we know what to make and we know how to adjust it, and the same principle applies in Harrison's case. As has been shown, the adjustments are not difficult. The hill test, by definition, must only take fifteen or twenty minutes, and as for the clock itself, yes, it has a remontoire which involves extra work in manufacture, but it is a good design and once made has not proved to be problematic. As for the cost, there are fewer parts in a Harrison-type regulator than the average contemporary English chiming clock, and its relative cost need not have been excessive. Also, most importantly, with the Harrison-type clock, which is designed to run without any lubrication, it needs little servicing, so both maintenance costs and the risk of damage from inexpert repairmen are largely avoided.

Harrison's original intention was that his regulator should provide very long-term accurate timekeeping for our very own institution, the Royal Observatory Greenwich. It is sad the development of this very special horological technology was not encouraged and taken advantage of in the mid-eighteenth century, when it could have been of real benefit to science.

References

Bateman, D. (2017). The antidote to John Harrison. *Horological Journal*, January, 24–35.

Harrison, J. (1763). *An explanation of my watch.* Available at https://ahsoc.contentfiles. net/media/assets/file/Explaining_My_Watch.pdf. Accessed 4 December 2018.

Harrison, J. (1775). *A Description Concerning Such Mechanism.* London, Jones, T. Available at https://ahsoc.contentfiles.net/media/assets/file/Concerning_ Such_Mechanism.pdf. Accessed 12 December 2018.

8

Crunching the Numbers

Analysis of Clock B's Performance at Greenwich

Tom Van Baak

Introduction

(1) In this chapter, measurement data for Clock B is examined in great detail. It confirms the success of barometric (air density) compensation in 2014, that the clock clearly met Harrison's 1 second in 100 days claim, and that further temperature compensation was subsequently required. Methods to claim accuracy are explored, the electronic measurement system that was used to monitor Clock B is described, and the various anomalies in raw data are investigated and explained. Finally, suggestions for subsequent Clock B experiments and advice to designers of precision pendulum clock timing systems are given.

(2) The detailed analysis of the accuracy of Clock B presented in this chapter is based on two sets of computer records. It is worth noting that, at the time, the author was not aware where the data came from or even what kind of clock was being measured. Only much later was the design, history, and location of the clock revealed. While this added to the challenge of analysing the data, it also greatly increased the objectivity of the analysis. The author works daily with mechanical, quartz, and atomic clocks and is impartial to the background of Clock B.

(3) One second in 100 days:

> there must be then more Reason (and that withal, as from Experience in my other Clock) that it shall perform to a second in 100 Days, yea, I say, more Reason, than that Mr Graham's should perform to a second in 1. (Harrison, 1775:35–36)

Van Baak, T., *Crunching the Numbers: Analysis of Clock B's Performance at Greenwich* In: *Harrison Decoded: Towards a Perfect Pendulum Clock.* Edited by Rory McEvoy and Jonathan Betts, Oxford University Press (2020). © Oxford University Press. DOI: 10.1093/oso/9780198816812.003.0008

What did Harrison actually mean when he made this boast? It is useful to imagine the many ways to claim an accuracy of 1 second over 100 days:

1. The simplest interpretation would be to set a clock to the correct time, lock the room, and come back 100 days later to see whether the error was within 1 second.

2. But what if the clock wanders outside the limits during the test? A fairer test (and likely the method that Harrison used) would be to set a clock to the correct time and check the error frequently. The error must be less than ±1 second; that is, the clock must never exceed 1 second ahead or 1 second behind at any time during the entire run, not just at the end of the run.

3. An even more stringent test would be to set the clock on-time, check the error frequently over 100 days, and require that the most positive error and most negative error differ by less than 1 second. This is known as the *peak-to-peak* error.

4. A more generous method would be to monitor the error each day and require only that the *average* error over 100 days is less than 1 second. Or one could use the *absolute* error each day so that positive errors don't favourably cancel negative errors. This is known as mean absolute deviation (MAD), an averaging method often used with time series data. Methods based on averages tend to be more forgiving of bad days and outliers.

5. An even more generous method is to collect 100 days of data and apply a retroactively calculated rate adjustment such that the total error is minimized. This is reminiscent of how chronometers are rated. It was also how the data from Shortt number 41 was analysed (Woodward, 1995:135–138).

6. One might even be allowed to start the clock, objectively collect data for a long time, and after reviewing the data, subjectively select the optimal start date such that the next 100 days remain within the one-second goal.

7. An engineering method would be to compute the root-mean-squared (RMS) or the standard deviation (STDEV) of all error measurements. These statistics are similar to MAD, but they penalize outliers quadratically. Note that ±1 second RMS over 100 days is a useful and valid measure of clock performance, but it tends to be a far easier goal to meet than a 1-second peak-to-peak.

8. A time and frequency metrology method would be to compute Allan deviation (ADEV) or time deviation (TDEV). These specialized statistics are used to measure rate stability in clocks and oscillators as a function of sampling intervals. Although very useful in many applications, their complexity, their requirement of many consecutive runs, and their forgiving treatment of clock error and rate are likely not what Harrison had in mind when he made his boast.

The approach taken with Clock B is one of the most conservative and difficult tests possible: option 2. That is, after some prolonged observation time, rate and time were adjusted and the clock case was shut and sealed; no rate adjustment was made to the clock or to any measurement data; the start date and run duration were announced ahead of time; and no clever statistics are used.

The goal was an error of less than one-second deviation in either direction from 0 UTC, over the first 100 days of 2015. Note that the 100th day coincided with the long-planned *Harrison Decoded Conference* in April 2015. To the author this seemed like a bold move: to announce a future conference to celebrate a result which was not measured yet. But the quality of early Clock B data supports the high confidence placed in the performance of the clock. And the clock did not disappoint.

Results for 2012–2013 (110 days)

Most of the charts that follow are based on two separate uninterrupted runs, the first in 2012–2013 and the second in 2014–2015.

Figure 8.1a shows the long-term performance of Clock B during an early run in 2012–2013. Data were plotted hourly from 31 October 2012 to 18 February 2013 for a total of 110 days. The graph shows the typical meandering error of a good clock keeping time to within a few seconds over the span of several months.

To be more specific, assuming the clock was on time at the beginning of the run, the peak-to-peak error is +1.28 seconds and −0.36 seconds; that is, clock error varied as much as 1.64 seconds during the run. The MAD is 0.58 seconds and the STDEV is 0.37 seconds. The rate stability or ADEV for tau 1 day is 5.76×10^{-7} or 0.050 seconds/day.

Although this level of performance for a pendulum clock is superb, one could argue that, strictly speaking, it does not meet the *1 second in 100 days*

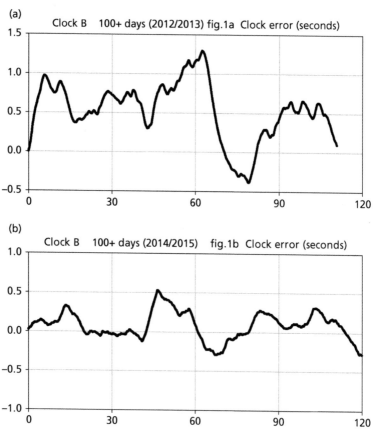

Fig. 8.1 (a) Clock B accuracy during early run, 2012–2013. (b) Clock B accuracy during later run, 2014–2015. Chart scale is fixed at 2 seconds of error high by 120 days of run time wide. © Tom Van Baak

requirement—because for several weeks the clock error exceeded 1.0 seconds from the starting point. Note that if the starting point were retroactively moved by ½ second, the clock might then claim ±1 second accuracy, that is, plus or minus 1 second from the adjusted mean in 100 days. In other words, if the trial were considered as having started 3 days into the run as shown, the clock would have met the requirement of 1 second in 100 days.

As will be discussed later, the clock had good temperature compensation but was in need of adjustment of its barometric (air density)

compensation during this run, which suggested that better performance was possible. Nevertheless, even in this less than optimal state,
performance was quite close to the goal.

Results for 2014–2015 (130 days)

Air density compensation adjustments were made in 2014. A second and
later run is shown in Fig. 8.1b. The data were collected from 1 December
2014 to 9 April 2015 for a total of 130 days. This plot includes the 100 days
of the formal Guinness record trial which began 1 January 2015. For easy
comparison, both charts (Figs. 8.1a 8.1b) use an identical scale: two
seconds of clock error vertical and 120 days of elapsed time horizontal.

Again, assuming the clock was on time at the beginning of the run,
the peak-to-peak error is +0.46 seconds and −0.34 seconds; that is, clock
error varied within a 0.80 second range. The MAD is 0.13 seconds and
the standard deviation (STDEV) is 0.17 seconds. The rate stability or
Allan deviation (ADEV) for tau 1 day is $2.92{\times}10^{-7}$ or 0.025 seconds/day.

Compared to the earlier run, the performance improved by a factor
of two and this later run meets a *1 second in 100 days* goal, no matter how
strictly it is defined.

Measurement system

Measurements of Clock B were made with MicroSet, a commercial
watch and clock timing system well respected and widely used within
the horological community. The system incorporates a non-contact
optoelectronic sensor, a microprocessor-based digital timer, an internal
quartz oscillator timebase and external GPS-based time synchroniza-
tion, an environmental sensor for recording temperature, humidity,
and barometric pressure, along with Windows PC software for serial/
USB data collection, archiving, and display.

The key elements in this measurement system are the optical sensor
and the use of GPS (or other UTC-locked standard) for long-term
accuracy. The infrared sensor for Clock B is seen in Fig. 8.2. For precision clocks this below-bob, flying-vane, photo-interrupter technique is
generally more reliable than video, mechanical, acoustic, or magnetic
sensors.

The infrared emitter/sensor elements are built into the clock providing maximum rigidity and precision. Oscilloscope measurements of the

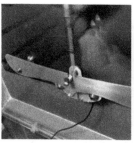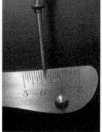

Fig. 8.2 For precision clocks this below-bob, flying-vane, photo-interrupter technique is generally more reliable than video, mechanical, acoustic, or magnetic sensors. (Left) The pendulum rate sensor is mounted at the centre of the beat-scale plate at the base of the clock and is solidly fixed to the back plate. (Centre) The rod extends ~10 to 15 cm below the bob and is threaded to accommodate a light weight rating nut. The final few millimetres are a thin vane which interrupts the optical sensor once per swing. Peak semi-arc is just over 6°. (Right) The sensor assembly consists of a clear plastic infrared (IR) emitter facing foreword and a dark plastic high-speed IR phototransistor facing backward. The interrupting vane and IR beam are ~1 to 2 mm in diameter, creating a narrow pulse approximately 2.3 ms wide. The leading edge is used for clock timing. (Photos by the author. © Tom Van Baak)

electrical signal from the sensor suggest it is the 'premium sensor' sold with the MicroSet. Rise and fall times are less than 1 microsecond (0.000001 seconds). The author observed that an external desk lamp continuously illuminated Clock B's sensor so that the ambient light level seen by the sensor was effectively immune to daily variations in daylight.

The GPS timing pulse was obtained with a Garmin 18x LVC, a compact GPS receiver with an output of one-pulse-per-second (1PPS). This model is a well-known and highly regarded GPS receiver and its precise electrical timing pulse is aligned to Coordinated Universal Time (UTC) to better than 1 microsecond.

The MicroSet timer was configured to record a Clock B measurement every 5 minutes. Unfortunately, plots often contained distracting artefacts and the raw data proved difficult to process due to numerous glitches. Over the span of 100 days one would hope for exactly 28,800 clean data points. At least 95 per cent of the data was clean but for a variety of reasons hundreds, even thousands, of data points were suspect. Was there something wrong with Clock B or something amiss with the measurements?

To validate the MicroSet system and to uncover the true cause of the anomalies, the author installed an independent measurement system during the latter part of 2015. This system was based on picPET, a hardware chip specifically designed for high-performance glitch-free timing. A series of simultaneous tests identified occasional problems with GPS reception in the horology workshop itself, occasional missed beats in the measurement system due to electrical noise from mains power, occasional jumping from measuring left swings versus right swings, and durations of corrupted data when the pendulum tick was too coincident with a UTC tick. All these anomalies were carefully analysed and resolved in the hourly data used for this report. Fortunately, and ironically, the glitches were due to modern technology and not Clock B.

2012–2013 data (110 days)

As mentioned earlier, both clock and environmental data were collected. Clocks do not change rate or jump in time by accident: at their core, pendulum clocks are sensors and their rate is affected by physical changes inside and out. The 2012–2013 data suggest Clock B was an excellent barometer. The 2014–2015 data suggest Clock B was a good thermometer. Perhaps at some point both the barometric and temperature compensation will be adjusted to a finer degree and then the true performance of Clock B will be revealed.

Figure 8.3 shows all the clock and environmental data for the early run in 2012–2013. Figures 8.3a and 8.3b show clock error and hourly mean rate as measured by MicroSet. Figures 8.3c, 8.3d, and 8.3e show ambient pressure, temperature, and humidity, respectively.

Figure 8.3f shows the variation in pendulum kinetic energy. The vane rapidly passes through the sensor at center swing and in addition to timing the pendulum, MicroSet precisely measures the duration of the interrupted beam. For Clock B the value is typically 2.3 milliseconds. With proper physical calibration constants one could translate this timing value into a measure of the pendulum's peak velocity in units of metres/second or perhaps peak amplitude in degrees. Even easier, and equally as informative, is to use relative units and display the result as a per cent variation in kinetic energy. This value can be useful for exploring the dynamic stability of the pendulum. It is not explored further in this report. Looking at Figure 8.3 one is immediately struck with the visual correlation between clock rate and barometric pressure. They are almost mirror images.

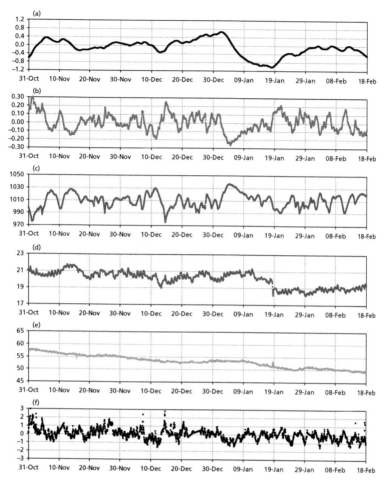

Fig. 8.3 Details for the 2012–2013 run lasting 110 days. (a) Clock B error (seconds), (b) hourly rate (seconds/day), (c) barometric pressure (mbar), (d) temperature (Celsius), (e) relative humidity (per cent), and (f) kinetic energy variation (per cent). © Tom Van Baak

2014–2015 data (130 days)

As presented earlier in this volume, the compensation was adjusted during 2014. The second run shows the result (Figure 8.4). Since the same scale is used for these six charts, they can be compared side by side. As hoped, the barometric pressure correlation dropped from 96 to

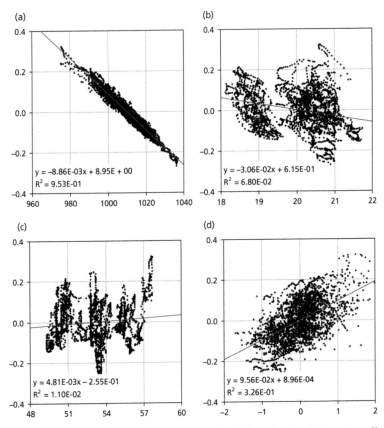

Fig. 8.4 Correlation scatter plots for the 2012–2013 run lasting 110 days. For all plots the y-axis is rate (seconds/day). Left to right, top to bottom, the x-axis is (a) barometric pressure, (b) temperature, (c) relative humidity, and (d) energy. Note the correlation (R^2) between barometric pressure and rate is extremely high: 95% © Tom Van Baak

4 per cent, meaning nearly full compensation had occurred. Unfortunately, temperature appears to have become more significant. It is evidently difficult to fix one without affecting the other. Nevertheless, the performance of the later run was dramatically better than the first.

Correlation and compensation

It is relatively straightforward, using visual analysis, to compare plots of environment and rate and from that make an educated guess on the direction and magnitude of needed adjustment. Or, as seen earlier, one can use simple statistics to obtain a more exact numerical value. In either case the normal course of action is to make informed adjustments to the compensation mechanism of the clock and then start another run to see whether performance improves. Sometimes this is an iterative process and multiple runs and multiple adjustments occur until no further gain seems possible.

The art of compensation is highly developed in modern electronic timekeeping. But rather than making mechanical corrections it is common to simply correct clock readings numerically rather than physically. There is no need for multiple runs in this case; one simply tries different compensation algorithms using existing data. Not only is this much faster, but it tends to be more precise.

Figure 8.5 shows how this technique can be applied to Clock B and the results are dramatic. We start with the graph in Fig. 8.5a, which shows the actual measured error in Clock B over the span of 110 days. From that, Figure 8.5b shows the actual hourly mean rate of the clock. Along with that, Figure 8.5c shows the actual recorded barometric pressure. So far, this is no different from other charts presented earlier.

From the correlation plot of Figure 8.6a we know the barometric correlation is 0.00886 seconds/day per mbar (equivalent to 0.103 ppm/mbar). Then, for each hour of real rate data, we calculate the portion of that rate that is caused by barometric pressure.

The assumption is that during any hour the rate is likely to be non-zero; clocks are not perfect. Some of that rate is likely due to temperature alone. Some of that rate is likely due to pressure alone and some may be due to changing air density caused by both. The rest of the rate is due to unknown effects; moreover, each clock has an inherent variation in rate. We can externally measure temperature and pressure. And we can compute the correlation coefficients. Therefore, those portions of rate can be removed.

It is a guess, an approximation. But the purpose of making the correlation charts is to show how good a guess it will be. In the case of 2012–2013 Clock B data, the guess about the effect of pressure is a near certainty. With 95 per cent confidence we know what effect a change in pressure will have on a change in rate. And so it is possible to take the pressure measurement, compute the effect on rate, and then subtract that from the measured rate.

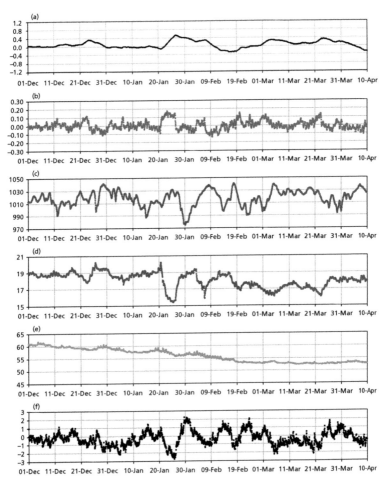

Fig. 8.5 Details for the 2012–2013 run lasting 110 days showing progression from real clock data to ex post facto compensated clock data. (a) Clock B error (seconds), (b) hourly rate (seconds/day), (c) barometric pressure (mbar), (d) hourly rate after numerical compensation, (e) time error after numerical compensation, and (f) chart scale to ±0.3 seconds. The 110-day performance of the software compensated paper clock shows improvement by a factor of 5: from 2 seconds to under 0.4 seconds. © Tom Van Baak

And then we subtract that from the real rate to create a virtual rate. A virtual rate is the rate the clock would have had if the barometric pressure effect did not exist. Since we know the real rate, and the real pressure, and the actual coefficient, we are able to calculate what the

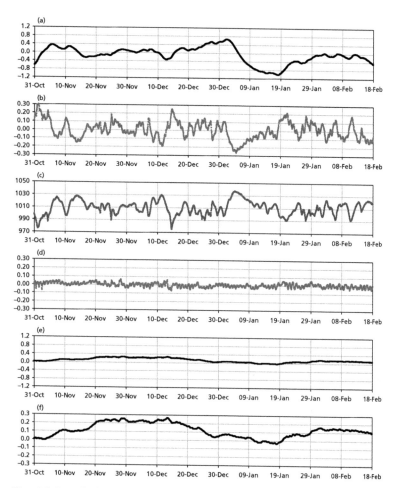

Fig. 8.6 Details for the 2014–2015 run lasting 130 days. (a) Clock B error (seconds), (b) hourly rate (seconds/day), (c) barometric pressure (mbar), (d) temperature (Celsius), (e) relative humidity (per cent), and (f) kinetic energy variation (per cent). Timekeeping performance is twice as good as the 2012–2013 run. © Tom Van Baak

rate would have been. This is done for every hour of data and the resulting virtual rate chart is shown in Fig. 8.5d. Calculations such as this have been done for centuries. Old pendulum clock literature also includes such methods as removing ex post facto certain measured physical effects. The concept of measuring time using these corrections is called a paper clock. And for work in astronomy there is no difference between using a real clock and a paper clock: whichever gives the best results. The paper clock by design gives better results.

Comparing the actual Clock B rate in Fig. 8.5b with the paper clock rate in Fig. 8.5d shows what a dramatic effect this has. Finally, converting paper clock rate to paper clock time is seen in Fig. 8.5e. Compare the actual clock error in Fig. 8.5a with the paper clock error in Fig. 8.5e.

In other words, if Clock B were used by astronomers in 2013 and they noticed that the clock was in need of barometric compensation adjustment they would have a choice. They could either stop the clock, make adjustment hoping for better results, or leave the clock running exactly as it is, and make the adjustments on paper. Clearly this is not what Harrison had in mind. But seeing the dramatic improvement in performance one can understand why paper clocks are used in modern timekeeping. For example, UTC, itself, is a paper clock.

Even if a paper clock is not used, the charts of Fig. 8.5 show what maximum potential that compensation adjustments can have, and they show the peak potential of the clock. In that respect we see in Fig. 8.5e that Clock B is performing to better than 0.3 seconds over 100 days, even better than Harrison imagined. In other words, doing the calculations to create a paper clock shows the clockmaker just how much effect, if any, mechanical adjustments to compensation can have. If the paper clock is the same as the real clock, then an adjustment is not advised. It cannot improve the clock.

Given that physical adjustments tend to take time, computing correlation coefficients, looking at R^2 values, and comparing paper clock performance with real clock performance can produce large time savings. The technique of paper clock or ex post facto software compensated timekeeping is not just for electronic clocks but is useful for mechanical clocks as well.

So how good would Clock B be if it were in a constant pressure room? One answer would be to do the experiment. Given the mass and size of Clock B this would be a difficult experiment. The other answer is to look at Fig. 8.5 and see that it would likely be five times better; from about ±2 seconds to ±0.4 seconds.

Astute readers will also note that one could apply paper clock corrections for both pressure and temperature and further improve the performance. There are a number of software tricks one can employ to extract the maximum possible timekeeping out of a clock; the potential performance of a clock is almost always better than the actual performance. For Clock B this technique is useful to precisely show the potential benefit of compensation before it is even implemented. Between correlation plots and paper clocks one can not only assess the effort-to-result ratio but also check how well the pre- and post-compensation live up to expectations.

Still, this is not what Harrison intended, and this section is not intended to make the clock look better than it is. But it is a technique that shows the potential in the clock, and that is very informative. A clock at its limit of potential needs no further adjustment. A clock running below its potential is worth improving.

High-resolution measurements

For Clock B measurements, the MicroSet timer is configured to collect data every 5 minutes. To verify that its readings were correct a different clock measurement system was installed in parallel to MicroSet during late 2015 using the same sensor signal—but different electronics, time-base, GPS, computer, and software. In addition, it was able to record every beat of Clock B to sub-microsecond accuracy. Although this created a hundredfold more data it provided a microscopic view into the operation of both Clock B and MicroSet.

One motivation for this test was to explain why Clock B appeared to be wandering in a 20-minute cycle. The charts in Fig. 8.7 explore this observed effect. The y-axis of all plots in Fig. 8.7 are seconds of error. The x-axis is elapsed time. Figure 8.7a shows MicroSet data for one typical day. Figure 8.7b shows picPET data for the same day. Clock B is wandering by about 15 milliseconds over a day. The two systems agree very well at this scale, which is good news. But when two systems agree so well, the next step is to zoom in for a closer look.

Figure 8.7c shows MicroSet data for 2 hours of that day. Figure 8.7d shows picPET data for the same 2 hours. Here a significant difference is evident. The MicroSet data show a roughly sinusoidal pattern with 6 cycles over 2 hours, or 1 cycle per 20 minutes. The picPET data shows a finer sinusoidal pattern with 30 cycles over 2 hours, or 1 cycle every 4 minutes. Figure 8.7e shows both data sets superimposed and this explains the mystery of the 20-minute wobble. A strong 4-minute

(a)

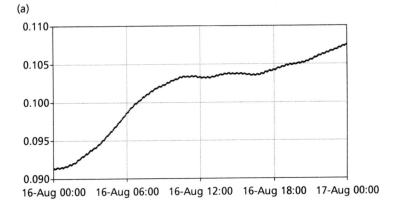

(b)

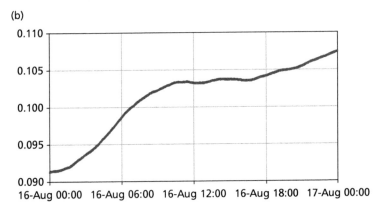

(c)

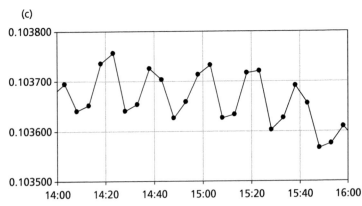

Fig. 8.7 Simultaneous measurement of Clock B by two different timers. The y-axis is seconds. (a) MicroSet over 1 day, (b) picPET over same day and matches perfectly, (c) MicroSet over 2 hours, (d) picPET over same 2 hours, (e) overlay showing data match and sampling mismatch, (f) high-resolution sampling over single 20-minute interval shows 30-second small variations within 4-minute large variations. © Tom Van Baak

(d)

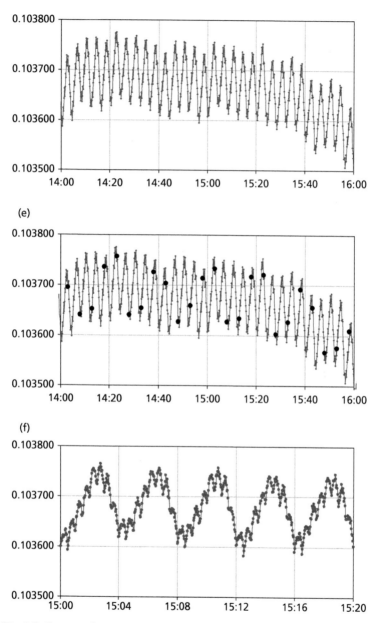

(e)

(f)

Fig. 8.7 Continued

variation in rate is present in Clock B. The MicroSet timer samples once every 5 minutes. Thus MicroSet captures the timing error of Clock B at different points along Clock B's 4-minute cycle, resulting in a 4 × 5 = 20-minute fictitious cycle. The unintentional sampling mismatch creates a slow measurement *beat note*, an artefact that isn't physically present in Clock B; a textbook example of aliasing and Nyquist–Shannon sampling theory: a Moiré pattern.

The 20-minute view in Fig. 8.7f shows clearly the 4-minute cycles in Clock B. Looking closer one also suspects there are smaller cycles within these larger cycles. By averaging multiple cycles together, the finer detail becomes visible. Figure 8.8 shows a single 4-minute cycle with clarity. Within that cycle are eight smaller 30-second cycles. At this level of detail even tiny imperfections in design and fabrication

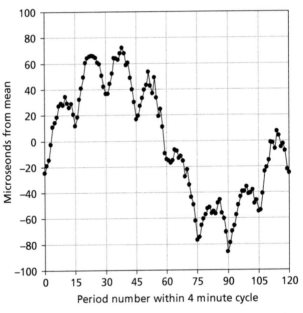

Fig. 8.8 Clock B has microscopic periodic rate variation in cycles of 30 and 240 seconds (4 minutes). Using 2-second picPET data, this plot was created by averaging many adjacent 4-minute segments together. Although each 4-minute cycle is very similar to every other 4-minute cycle, the eight 30-second cycles within each 4-minute cycle are significantly different. It is not clear whether this is intentional, or whether it affects timekeeping. Erratic but periodic variations tend to average out in the long term. © Tom Van Baak

become evident. What effect this has on long-term timekeeping is not clear. It certainly contributes to the short-term variations in the time shown by Clock B, but much of this variation appears to average away over time.

If nothing else it demonstrates that measuring a pendulum clock at this level of precision easily reveals imperfections; understanding imperfections of any kind is a necessary step toward improving clock performance. The existence of the 30-second, 4-minute, and accidental 20-minute beat note cycles and occasional missed samples is one reason why the Clock B data were analysed in this report using hourly averages. At that level cyclical effects are reduced to a minimum and the data are cleaner.

Results, long term (476 days)

Clocks are complex objects and measurement of time is often more complex than other physical measurements. The expression '1 second in 100 days' is a dimensionless ratio, 1 s/100 days, equivalent to 1 s/8,640,000 s, which is 0.115 ppm (parts per million), or 1.15×10^{-7}, or 115 ppb (parts per billion).

Still, while 1 second in 100 days is *numerically* equivalent to 0.1 seconds in 10 days or 0.01 seconds a day, they are not at all equivalent *horologically*. It tends to be far more difficult to achieve 1 s/100 days than it would be to achieve 0.01 s/day. The reason is simple: clocks are complex real-world sensors and what is measured as time error is simply the sum total of a large number of large and small, fast and slow, random and systematic time-varying effects. In general they do not magically *average out* over time. Accurate timekeeping does not just get harder over time; it gets increasingly harder over time.

The fact that the clock error and rate are moving targets while measurements are being performed has ramifications on the statistics used to analyse the results. Conventional thoughts like 'it will get better over time' or 'it will average out' rarely apply to clocks. In fact, over the long term, for many clocks, the more one averages, the worse the accuracy is. Figure 8.9 shows the results for the longest continuous data set available. At about 16 months it is more than four 100-day runs back to back. The results are impressive, but just because a clock achieves 1 s/100 days once does not mean that it will continue to be within 1 second for any 100 days that follow.

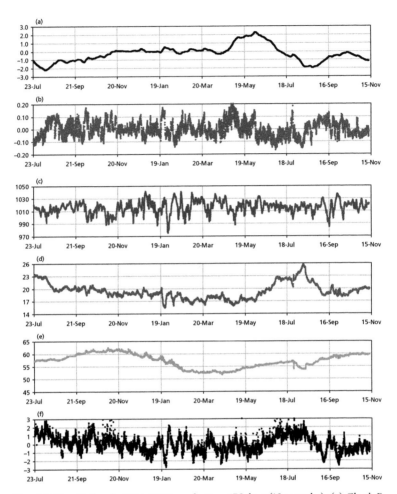

Fig. 8.9 Details for the 2014–2015 run lasting 476 days (16 months). (a) Clock B error (seconds), (b) hourly rate (seconds/day), (c) barometric pressure (mbar), (d) temperature (Celsius), (e) relative humidity (per cent), and (f) kinetic energy variation (per cent). Timekeeping is within ±2 seconds. © Tom Van Baak

At the same time, it can be seen in retrospect that if the Guinness trial had started in April 2015 instead of in January 2015, the clock would not have met its goal. Thus, there is a certain element of serendipity at work. To guard against this to some extent, one might specify a condition such as '1 second in a 100 days, during each of four seasons, over the span of a year'.

Leap second

It is well known that a year, as measured by Earth's revolution around the Sun, does not contain an exact integer multiple of days, as measured by Earth's rotation about its axis. A solution is a *leap day*—where approximately every four years one day is inserted into the calendar.

Similarly, but less well known, is that the Earth day, as measured by Earth's rotation upon its axis, is not an exact integer multiple of atomic seconds, as defined by atomic frequency standards. A solution is a *leap second*—where one second is inserted into (or deleted from) UTC. Interestingly, there was a leap second in 2015 and it affected Clock B measurement data. Near the end of 30 June 2015 Clock B error was +0.66 seconds (Clock B ahead of UTC). One second later, near the beginning of 1 July 2015 Clock B error was −0.34 seconds (Clock B behind UTC). That had the unintentional effect of reducing the apparent error of Clock B by 0.3 seconds!

However, since Clock B was rated for UTC, and continuously measured against UTC (via telephone, radio, internet, and GPS), and leap seconds are unpredictable, leap seconds should not be included in the measurement of a pendulum clock. When measuring the performance of a clock, one should always use a *reference* clock that is a significantly better timekeeper. It is remarkable that the best pendulum clocks must be measured by an *atomic* clock and not an *earth* clock.

Clock B suggestions

The data presented answer the Clock B accuracy question. But to further understand how, or how well, the clock actually works it would be useful to study it under different conditions.

- How does the performance of the clock depend on the mass of the wall?
- How does the clock respond to tilt, vibration, or seismic perturbation?
- How well would Clock B perform in a cellar with 1 or 0.1°C temperature stability?
- Would clones (Clock C?) give the same performance as Clock B?
- How consistent is the rate if the clock were stopped and restarted daily?

- How quickly does the clock re-establish its previous rate after a disturbance?
- How would it react to a forced 1, 3, 10, or 30 per cent increase or decrease in amplitude?
- Is Clock B so accurate that it can detect the effect of lunar/solar tides?
- What mechanical resonances are present in different parts of the clock?
- How well would Clock B perform over several years or a decade?
- Are there any components whose changing state would cause a gradual drift in the clock's rate over years?

Finally, if possible, a similar series of tests should be applied to other high-quality pendulum clocks which claim performance near 1 second in 100 days. There is not likely time or motivation to perform all these tests on a variety of clocks, but it should be our goal that careful scientific measurements replace heated subjective arguments.

Other questions may not be answerable with science alone. For example, can one fairly compare the performance of two clocks, one with wheels and hands and the other a tank regulator with electronic slave? Can one fairly compare two clocks when one is exposed to an office environment and the other is deep in a specially designed underground chamber?

Measurement suggestions

A wide variety of measurement systems exists for clocks and watches. But a world-class precision pendulum clock presents a challenge that few measurement systems are prepared for. During the several years of measurement of Clock B there were problems with workshop power, with Windows software, with GPS reception, with missed samples, and with corrupted samples. In every case, the problem was due to the workshop and measurement system and not the clock itself.

Fortunately, none of the errors impacted the quality of the data, long term. In other words, the glitches in the data made the work of analysing the raw data inconvenient, but did not impact the final outcome of the result.

Here are some suggestions for those making measurements of precision pendulum clocks:

- Use high-quality optical sensor and signal conditioning electronics;
- Use oscilloscope to verify microsecond-level rise and fall times;
- Minimize sources of timing jitter and sensor direction asymmetry;
- Minimize the effect of ambient lighting and temperature on sensor;
- Make the sensor an integral part of the pendulum construction;
- Use shielded cables and pay attention to earth grounding issues;
- Try to make and record measurements at the microsecond level;
- Use continuous timestamp method instead of period or time interval;
- Record both the time of the pulse and the pulse width, or simply timestamp both edges of the optical pulse;
- Use low-drift ovenized quartz, atomic, or GPS-locked timebase;
- Use reliable or redundant sources of UTC and test them in place;
- Do not record data at a rate inconsistent with wheel cycles;
- Record as much data as possible, even as much as every beat;
- Record data to PC or storage media, independent of GUI software;
- Isolate the PC from the Internet and disable background tasks;
- Eliminate arbitrary software limitations on run time or data size;
- Use battery backup on all electronics, whether 5 volts or mains;
- Environmental data should be recorded in parallel with clock data;
- Employ multiple temperature sensors for thermal gradient studies; and
- Identify and solve anomalies immediately rather than years later.

Conclusion

The measurement system used in the Greenwich workshop employed many of these features and practices, but not all. As a result thousands of data points were missing or unusable during each year, and failure of the PC's power supply in December 2015 ended data collection in the Horology Workshops. However, because the Microset measurements were regulated by GPS and the clock was consistently measured every 5 minutes on average, this had negligible impact on the fidelity of the hourly graphs presented in this report.

The author wishes to thank Donald Saff (Clock B benefactor), Jonathan Betts and Rory McEvoy (Greenwich curators), Bryan Mumford (MicroSet designer), and Bob Holmstrom (*Horological Science Journal*) for their invaluable assistance in the data analysis and creation of this report. It is hoped that the data collected and analysed will help with the understanding of this unusual pendulum clock.

References

Harrison, J. (1775). *A Description Concerning Such Mechanism.* London, Jones, T. Available at https://ahsoc.contentfiles.net/media/assets/file/Concerning_Such_Mechanism.pdf. Accessed 12 December 2018.

Van Baak, T. (2016). *A technical guide to Clock B raw data.* Available at http://www.leapsecond.com/pend/clockb/index.htm

Woodward, P. (1995). *My Own Right Time.* Oxford, Oxford University Press.

9

Decoding the Physical Theory of Harrison's Timekeepers

M. K. Hobden

Harrison (1775:23) stated: 'Mr Graham said to several Gentleman, that for my Improvement in Clock-Work, I deserved 20000 [pounds]—was no Longitude to be concerned'. The running of his early precision clocks to a second a month gave him every confidence that his new regulator, when finished, would easily fulfil his prediction of a second in 100 days. In his early years, Harrison's isolation from mainstream horological developments in London had forced him to depend on the available scientific literature of his day and his experimental experience working with his timekeepers.

As we shall show, from the physics propositions in Newton's *Principia* and Nicolas Saunderson's 'Mechanics', he was able to form a complete model of a self-excited oscillating system (a system which gains its energy of maintenance intermittently from a discontinuous source such as a falling weight or tension in a spring). He formed a comprehensive understanding of how it behaved under perturbation and, therefore, could predict the final performance. Saunderson, who was blind following a childhood illness, succeeded William Whiston as Lucasian Professor of Mathematics at Cambridge in 1711. Harrison is known to have had a copy of Saunderson's 'Mechanics' given to him by a visiting clergyman (Quill, 1966:14).

In analysing Harrison's clock system, it is necessary to use the mathematics of his era—synthetic geometry and fluxional calculus—not the later 'linear' algebras of the Continental school which failed to generate a proper theory of the self-excited oscillator. Modern commentators have tended to ignore the work of the British fluxional school, or they translate it into a 'modern' mathematical form which does not fully represent the thought processes of the original author at that period.

Hobden, M. K., *Decoding the Physical Theory of Harrison's Timekeepers* In: *Harrison Decoded: Towards a Perfect Pendulum Clock*. Edited by Rory McEvoy and Jonathan Betts, Oxford University Press (2020).
© Oxford University Press. DOI: 10.1093/oso/9780198816812.003.0009

Newtonian physics

Newton's *Principia Mathematica* of 1687 for the first time provided a firm basis both for understanding and extending knowledge of the physical world and for promulgating mathematical physics to a much wider audience. With the publication, on Edmund Halley's insistence in 1686, Newton was rapidly elevated to an almost superstar status across Europe, a position he was sometimes forced to defend in the coming decades, but it was the popularisation of his major work and its mathematical principles by mathematical practitioners which probably influenced Harrison most.

In *Principia*, Newton gives a very thorough account of the motion of pendulums in 'resistive media' (air and liquids, for example) which was essential to his experimental evaluation to prove the equivalence of inertial and gravitational mass. In doing so, he invented the first ever phase plane representation of an oscillating system (Fig. 9.1). To reduce

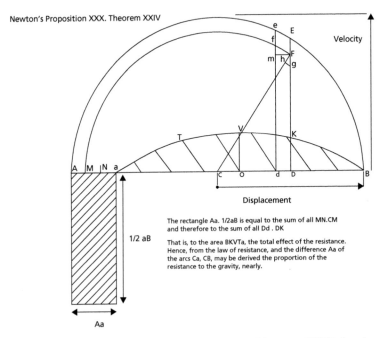

Fig. 9.1 Diagram from Newton's Proposition XXX, Theorum XXIV, showing the retardation of a pendulum, relative to the resistant medium it swings in (Newton, 2010:246). © Mervyn Hobden, 2001.

his experimental results, using the decay of pendulums of different weights and density, he had to have a comprehensive understanding of the air's resistance and how it affected the decay curve. In the following General Scholium, he describes the careful experiments he made to find the law of the air's resistance with amplitude. He proposed that the change in resistance follows a simple polynomial law proportional to $AV + BV^{3/2} + CV^2$ where V = the velocity of the pendulum bob at any point in the arc. Then, from the experimental results with different 'globes' or bobs, he calculates the coefficients for his equation and proves that it is consistent. This is not just a mathematical demonstration but the use of a consistent physical method to quantify the correctness of his analysis.

Harrison, using the demonstrations in his copy of Nicolas Saunderson's 'Mechanics', assumed correctly that the predominant term at his running arc is as V^2—it is doubtful he ever saw a copy of the *Principia* before he travelled down to London and it was not available in English, until Andrew Motte's 1727 translation—but certainly Halley or Graham would have made Harrison very aware of what it contained and its importance. Newton went on to show that changes in resistance acting on the pendulum can always be related to the changes in the arc of swing and their resultant 'fluent quantities' or what we would now call the resultant integrals of work. In Proposition XXX, using a phase plane model, he shows that the ratio between the work done by a linear resistance to the change in arc is as '11/7 very nearly' or as radius squared to the complete area of the semicircular phase plane. This ratio, effectively $\pi/2$, appears to be the origin of Harrison's ratio for the obliquity given to his input force as the pallets interchange—'as 3 is to 2' (Hobden, 2011:2–13).

In the next proposition, XXXI, Newton stated that:

> If the resistance made to an oscillating body in each of the proportional parts of the arcs described be augmented or diminished in a given ratio, the differences between the arc described in the descent and the arc described in the subsequent ascent will be augmented or diminished in the same ratio (Newton, 2010:248).

This is a clear statement that the changes in the arc are in the ratio of the difference between two quadratures, or integral quantities relating arc to resistance—in this case the difference in the total stored energy of the oscillating body to the work done by the resistance. Newton went on to show how this ratio changes, dependent on the law of resistance chosen. Harrison used exactly the same definition in words in his *Explanation of My Watch* for the Board of Longitude. A similar explanation was used by Professor David Robertson in his (1928–1932) series in the *Horological Journal*, which proved invaluable in clarifying Harrison's ideas.

Popularising/commercialising Newtonian physics

The act of dissemination was not just undertaken by universities and the Royal Society, but by a class of mathematical practitioners whose goal was the practical application of Newtonian physics and education. For many, demonstrating the solution of practical problems provided an income for them (Taylor, 1954).

This advanced rapidly in the seventeenth century with Hooke, Moxon, Mandey, and many others producing published works explaining mechanical and optical principles, all extremely attractive to a literate middle class. There was a flourishing trade in books and lectures, together with the foundation of learned societies to feed these new concepts to an eager audience, ready to take up this revolutionary take on the possibilities of science and its interaction. It was apparent that these new principles gave potential access to the levers of economic power. This was an age when formal education was confined largely to the classics and apart from a few notable exceptions, such as Isaac Barrow at Cambridge, little effort was made to incorporate 'science' into the curriculum.

Venteri Mandey garnered a comfortable income from translating these new principles into a practical form understandable by the audiences at one of the new corresponding societies which were responsible for local interest, such as Maurice Johnson's Spalding Society which had Newton as a corresponding member. The frontispiece from his book gives a very clear statement of his motives—this is the practical application of science for economic gain. Far from being a dry dissertation on bare mathematical principles this is about application: 'in removing and raising bodies of vast weights; with little strength or force and also the making of machines or engines, for raising of water, draining of grounds and several other uses.' He also dealt with the 'making of clockwork and other engines'. And concluded with the assurance that the contents are 'both pleasant and profitable for all sorts of men, from the highest to the lowest degree.' There is synergy between Mandey in 1702 and those earnest presenters of science seen on social media today.

> that the Length of a Pendulum, as at the best, is only as in Proportion to the Length of the Pallats (Harrison, 1775:55)

Mandey compared how the effects of a force, acting on a pendulum, vary according to the length of the pendulum, the mass of the bob, and the point of application of the force with respect to the pivot point (Fig. 9.2). This is an idea that resurfaced in Harrison's work with his insistence on the importance of the moment arm of the application of the driving force from the scape wheel tooth to the pendulum. He criticised George Graham's clock for the length of the pallet arms on the deadbeat escapement and stated that 'in my clock the pallats have not 1/4 part of the distance from its centre that Mr. Graham's small vibration is maintained.' And stated that as far as Graham's pallet arm length was concerned,

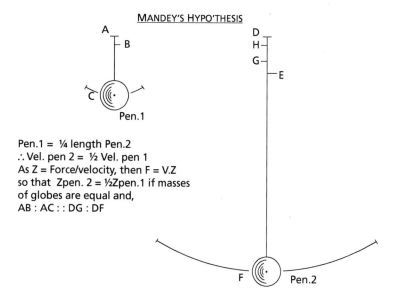

MANDEY'S HYPO'THESIS

Pen.1 = ¼ length Pen.2
∴ Vel. pen 2 = ½ Vel. pen 1
As Z = Force/velocity, then F = V.Z
so that Zpen. 2 = ½Zpen.1 if masses
of globes are equal and,
AB : AC : : DG : DF

Fig. 9.2 Venteri Mandey's hypothesis on the application points of maintaining force to a pendulum. © Mervyn Hobden.

Well, granting this to be or if it be an excellency then it must be better to enhance or augment it i.e. to carry it farther, and the farther or more so still the better, so it must be the best for the pallats to be at least as long as the pendulum i.e. in effect for the action of the pendulum wheel to be at or upon the bob itself... where is hardly the pendulum at all and be the bob ever so heavy? (Harrison, 1763)

Dominion

This is the idea of relative stored energy of motion 'vis viva' of swinging pendulums, as compared by the ratio of their relative lengths, bob mass, and their relative velocities and then factored by the point of application of the maintaining force. Harrison went on to claim that this allowed him to say that in his timekeeper, he has obtained 'the greatest vibrations...from the smallest or from a given force' (Harrison, 1763:16). This is equivalent to the modern idea of admittance, the ratio of velocity (or angular frequency) to the applied force. Harrison then explained his concept of 'dominion over the wheels of the clock':

a bob of three pound weight describing an arch of 12 degrees will be equal in power for the regulating a clock, as a bob of 48 pound describing an arch of 3 degrees (Harrison, 1775:41)

'Vis viva' or energy of motion is ∝ mass × arc squared—a factor of ½ and the semi-arc is used in modern mechanics. In Harrison's example, both give 432 'units' (Hobden, 1982). A similar calculation can be found in Harrison's 1730 manuscript—the idea of 'vis viva'—energy of motion was generated by Christiaan Huygens in the late seventeenth century. Harrison's 'Dominion over the wheels of the clock' is proportional to θ^2 ω^2m, where, θ = amplitude in radians, ω = angular frequency in radians per second (rad/s), and m = mass. The higher the stored energy and the more rapid the energy exchanges in the resonator, the more difficult for an external force to act to change its motion.

This understanding forced Harrison to change, from the early marine timekeepers with balances beating seconds, to the rapidly beating balance of the watch, H4, with 5 beats/s. The resistance to external change isn't simply proportional to the mass of the pendulum or balance—it is the moment of momentum, which in classical physics describes the potential for resisting change of a moving body, that gives the pendulum or balance its dominion over all contrary forces and which can be related directly to the stored energy of motion (Le Corbellier, 1960). The modern calculation gives 0.146 joules for Harrison's pendulums—this can be compared with the 0.007 joules of the Shortt free pendulum clock—21 times less stored energy to do work against any perturbing forces.

Lagrange and Laplace, who made no attempt solve the deep-rooted problems inherent in the Newtonian fluxional calculus, formulated instead a structure based on the concepts of linear algebra in which the problems could not exist, built around the concept of the linear oscillator generated by Leonhard Euler in 1739. This is exemplified by Airy's (1827) paper on escapements, designed to advertise the ideas of the Analytical Society, who were trying to remove the use of the Newtonian calculus by saying that it was 'well into its "dot-age!" '—this being an unkind reference to the dot notation used by Newton for differentials. In actual fact, considerable progress was made in in the use of the fluxional calculus in England and it made use of concepts that only reappeared in the twentieth century with the development of the ideas of nonlinear theory.

Dynamic stability

neither...does...the Air's Resistance, want to be avoided, as many have foolishly imagined, but is of real or great Use.

(HARRISON, 1775:27)

The most fundamental idea that Harrison produced is that of dynamic stability. This section of the chapter discusses the constituents and mechanics of Harrison's system to self-correct when perturbed and return to the previous steady state as quickly as possible and doing so with minimum accumulated error with respect to time. Linear oscillating systems are dynamically unstable. This is well understood by modern vibration engineers, and the first person after Harrison to investigate this was Lord Rayleigh, in the *Theory of Sound* (Strutt, 1883).

As this section shows, examining Rayleigh's doubts over the ability of a linear algebraic model of an oscillator to explain the physical existence of a steady state, Rayleigh proved that only a nonlinear model can explain what is observed in physical systems.

In the 1820s it was complained that, with their 'steady state' idea, Lagrange and Laplace had removed dynamics and replaced it with statics. This simplified the processes of differentiation and integration—motion in a resonator now being 'simple harmonic'—into reversible processes based on closed integral curves—a completely linear solution, but only valid for a conservative or lossless system. With any resistive term, there is no asymptotic stable closed solution (Strutt, 1883).

A cycloidal, lossless, linear pendulum takes up the amplitude set by the initial conditions—if not further perturbed, it remains at that amplitude in the phase plane. This is the assumed steady state. This is not the case with the self-excited oscillation of a nonlinear system. Rayleigh remarked, regarding his non-linear differential equation:

$$\ddot{x} + k\dot{x} + k'\dot{x}^3 + \omega^2 x = 0$$

If k and k' both be positive the vibration will die down, while if k and k' be both negative, the vibration will increase without limit. If k be negative and k' positive, the vibration becomes steady and assumes the amplitude defined by the first order solution. A smaller vibration increases up to this point and a larger vibration falls down to it. If on the other hand k be positive and k' negative, the steady vibration abstractly possible is unstable, a departure in any direction from the previous steady state amplitude tending always to increase.

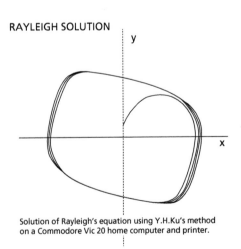

RAYLEIGH SOLUTION

y

x

Solution of Rayleigh's equation using Y.H.Ku's method
on a Commodore Vic 20 home computer and printer.

Fig. 9.3 Rayleigh's equation $\ddot{x} + k\dot{x} + k'\dot{x}^3 + \omega^2 x = 0$ in the phase plane.
© Mervyn Hobden.

—an unstable result where the coefficients of the damping terms have the wrong signs (Fig. 9.3). Stability to a steady state has to be designed in—it doesn't just happen!

Harrison discovered the limitations of stability on the Brocklesby Park stable clock—the vanes on the pendulum bob reduced the amplitude but increased the instability of the arc (Hobden, 2015). Harrison (1763) stated that the total arc must be limited to avoid instability due to the changing air's turbulence with increased velocity.

Figure 9.4 shows stability tests on Clock A at large arcs. The top trace is at 8° semi-arc; the bottom trace is with a card vane clipped to the bottom of the pendulum rod. The combination of increased drag and increased nonlinearity greatly increases the oscillator's noise level as well as dramatically altering the rate. Harrison was aware that the added noise from the air's motion would normally only be of short-term duration and this rapid convergence to a steady state, later confirmed by Allan variance measurements, allowed rapid adjustment of the clock (Jeans, 1925).

Rayleigh's is the first clear statement after Harrison that stability in an oscillator is conditional on the existence of the right ratio between

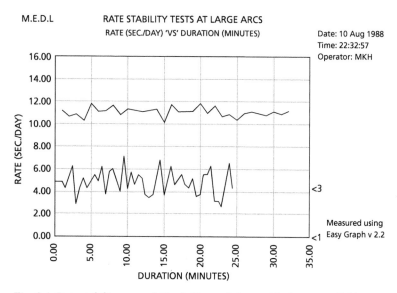

Fig. 9.4 Rate stability tests of Clock A's pendulum with damping. © Mervyn Hobden.

positive and negative resistance terms in the governing equation. This led, as later shown by Henri Poincaré, to a single closed curve that was later termed a 'limit cycle' in the solution, where the oscillator always returns to a defined amplitude and period. Rayleigh's nonlinear equation describes where that amplitude limiting occurs as a function of velocity. Harrison's solution contains both velocity-limiting and displacement-limiting terms.

The Rayleigh solution, whatever the initial conditions, converges to a parametrically defined steady state dependent on the values and signs of k and k'. This is exactly the behaviour seen by any clock or watch-maker: for any given set of parameters the amplitude converges to a single amplitude and period of oscillation. A displacement-limiting equation, equivalent to the linearised grasshopper, was generated in the 1920s by Balthazar van der Pol (Van der Pol, 1934):

$$\ddot{x} + \mu(1-x^2) + x = 0.$$

This is a normalised equation with displacement measured around the point where the negative and positive resistances cancel. The damping term in the brackets $(1 - x^2)$ is governed by μ, which is the input force/spring force ratio acting in the system (Fig. 9.5). Damping now varies as the displacement squared and, like the Rayleigh equation, asymptotes to a single closed curve dependant on μ. Harrison, having described the two linear possibilities of transferring energy in and out of a pendulum or other resonator and rejecting them as 'not to be done', went on to describe his reasoning behind the choice of the obliquity ratio for his maintaining force so as to provide isochronous performance:

> But now thirdly, and the which may be done (or is what I have done) viz the force of the wheel to be so disposed of by the construction of the pallets (and as thence without any sensible friction, or at least at all times without any sensible difference in the friction) so as a much greater impulse may be imprest upon the pendulum towards the lat-ter end of each, or every one of its ascents, than at the beginning of each, or every [one] of its descents, yea so much greater as to bear with, or truly to ballance the acceleration, or rather as more properly speak-ing the hastening arising from a proper recoiling, as must be when (and that in a right proportion) the said hastening power will thereby become less and less, the greater that space or quantity of the vibra-tion, as occupied in the said part of recoiling ever at any time happens

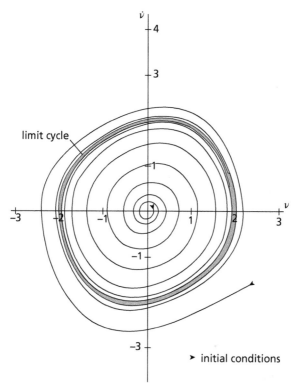

Fig. 9.5 Solution to the Van der Pol equation $\ddot{x} + \mu(1-x^2) + x = 0$ in the phase plane. As with the Rayleigh equation with the correct coefficients, it converges to a single closed integral curve. © Mervyn Hobden.

to be; for this (or these) as I found from experience (and as implied with the vibration large) can be so ordered as that truly to take in (whenever it may so happen) both a different force from the wheel and a different resistance from the air, as well as at the same time, to be truly adapted to either of which alone. And it is to be observed that if this be not the case; then (to the point of truth), no cycloid can take place. (Harrison, 1763:44–45)

Without such an offset force ratio, the recoiling action must always cause a shortening of the period of swing with increased force from the wheel—as was well recognised by both clockmakers and astronomers at this period (Fig. 9.6). To render the period isochronous, the ratio

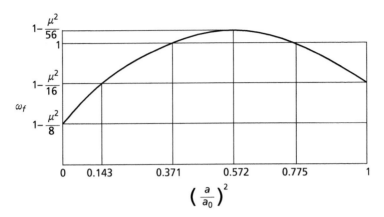

Fig. 9.6 Period hill in the Van der Pol equation, as predicted by Zia Acasu (1954) © Mervyn Hobden.

between the two quadratures must be a variable and this leads to the 'hill' in the equation solution.

The existence of the hill in the period response of a nonlinear oscillator was rediscovered in the twentieth century—it cannot be predicted from linear theory as the oscillator characteristic is assumed linear and infinite (Groszkowski, 1933). Martin Burgess, while testing Clock 'A' in 1979, became aware of the existence of a hill in the measured period function. Our research into the Van der Pol equation led to discovery of the work of Ziya Acasu, who predicted the existence of such a hill in the transient as the oscillator builds up to a steady state at its running amplitude (Acasu, 1954).

The other plank in Harrison's understanding was the nonlinearity of the pendulum—the period of a simple pendulum increases with increasing amplitude. This had been analysed by Christiaan Huygens in his *Horologium Oscillatorium* of 1673 where he shows that the path of the bob must follow a cycloidal path for the period to be isochronous. Harrison (1763) told us that he knew of this effect from 'being a Ringer' and observing that the period of a bell increases as the amplitude increases. He remarked:

> Now the Nature of such a Matter, or Cycloid to the Purpose, (and as consequently withal for preserving the Spring) must be as in some Measure reverse to what is demonstrated by Mr. Huygens, &c. that is, it must be

so as to occasion little Vibrations of the Pendulum, vis. all such as are less (and unregarded) than so as to let, or such as will let the Pallats interchange, to be still sooner performed, than what they would as otherwise be without it; and at such an Arch describing, as whereby just to let the Pallats interchange, or as rather at a little bigger, the Length of the Pendulum to be so [viz, as by or from its adjusting] as then to swing Seconds, and also, as when in its fetching farther, [as from the Nature of such a Cycloid as must be, and as when together upon such other Foundation as above described] the same; for as thence, from the Continuation of the circular Curvature of the Cheeks, (viz, of this artificial Cycloid) that Matter, as here in Hand, is to be ascertained. (Harrison, 1775:46)

The period of oscillation at which the pallats interchange should be slightly faster than the period at the running arc—exactly as predicted later by Acasu (1954) for the Van der Pol (VDP) equation as the point where the positive and negative resistances acting on the pendulum are exactly equal. As the arc increases to the running arc, the positive resistance due to the normalised recoil increases. The combination of the escapement VDP function with the spring and cheeks gives a slope which is substantially linear and which is required to compensate for the flotation effect caused by changes in air density, affected by changes in both air temperature and pressure, as demonstrated by Martin Burgess (1996).

Measured data show a combination of spring and circular arc cheeks does indeed give a flat-topped hill, as predicted earlier by Harrison with the linearized grasshopper escapement, with marked change in the rate of change of phase dependent on the spring thickness. The third plot (Fig. 9.7) is that of the phase change with amplitude of a non-compensated pendulum, which gives a close to sinusoidal function as the arc falls and the period decreases. We were not the first to reach such a conclusion: Smith (1964) demonstrated that restriction of the effect of nonlinear deviation of the pendulum period was not only possible but could be easily demonstrated to college students.

let us Suppose ye Clock Pendulum to describe any Certain Arch, & ye Air at any Certain Weight (Harrison, 1730:6)

Robert Boyle discovered in 1660, using Hooke's vacuum pump, that the decrement of a bob pendulum (the ratio of arc change to total remaining

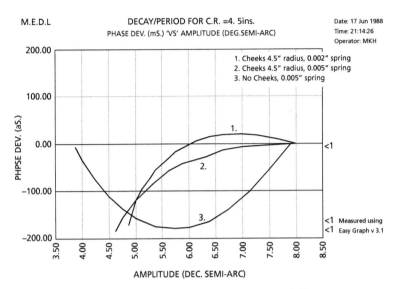

Fig. 9.7 Pendulum decay experiments on Clock A with different suspension spring thicknesses. © Mervyn Hobden.

arc) was largely independent of pressure. This caused considerable conflict with his simple model of the spring of the air. This was later confirmed by James Clerk Maxwell's study of the *Kinetic Theory of Gases* (1866). However, viscosity changes rapidly with temperature, so that the mean air resistance varies between the pressures in winter to those in summer.

> the Pendulum withal requiring to be (viz. as from my Contrivance of its Combination of Brass and Steel Wires) rather, as mathematically speaking, shorter when warm than when cold. (Harrison, 1775:25)

While the decrement of energy lost by the pendulum is largely independent of pressure, the extent of arc loss is principally governed by changes in kinematic viscosity of air, caused by temperature change.

Harrison would have been aware of Newton's Law of Resistance for fluids and gases from Sanderson's lectures—which we now call dynamic viscosity, and he was well aware of flotation on his compound gridiron pendulum—the air density flotation effect on the effective length of the pendulum 'with the Air at any Certain weight'—he knew this before 1730. It was well understood in the eighteenth century that mass above the centre of oscillation of a pendulum altered both the period and the flotation effect, and the distinction between 'simple' and 'compound' pendulums is clearly made from the publication of the *Horologium Oscillatorium* by Huygens in 1673 onwards.

The fact that the mean air density and its resistance vary in this way was explained by Professor David Robertson in a series of articles (1928–1932) that shows there is a fundamental difference between the performance of a pendulum sealed in a case with temperature and that of a pendulum in atmospheric air, in terms of arc change with temperature.

The kinetic theory of gases, where μ is the dynamic viscosity, ρ is the density, P is the pressure, v is the kinematic viscosity, R is the gas constant, and T is the temperature, shows that

$$\rho = P/RT, v = \mu/\rho, \text{and therefore } v = RT/P.$$

This gives us $\mu/v = P/RT$; i.e. the ratio of dynamic viscosity to the kinematic viscosity is equal to the pressure divided by the product of the gas constant and the absolute temperature. Density decreases as temperature rises but kinematic viscosity increases. So the arc change required to compensate for flotation at a given temperature cannot compensate exactly if the temperature rises. To compensate for the change in the ratio, the pendulum must be shorter when warm.

Harrison had observed the changing relationship between flotation and arc change as the air density and the kinematic viscosity changed with temperature and put in the necessary correction before 1730. He previously had found that, as was pointed out to me by Peter Hastings, using exact ratios of steel and brass leads to undercompensation in a compound (gridiron compensation) pendulum and therefore that 'Mr Graham had not the redoubling of his brass and steel wires...by a good deal, no, not even by 2 inches or more' (Harrison, 1775:12) to counteract this effect. The additional compensation for the residual temperature error, making the pendulum 'shorter when warm', would apply to the grasshopper escapement with circular arc cheeks correctly set for barometric compensation, not to a Graham regulator with dead-beat escapement.

The plot in Figure 9.8 shows the hill taken with the remontoire locked and run down to the point where one pallet was hanging on the scape wheel tooth, measured using the Beckman averaging counter. However, it shows the slope of the pendulum hill on that part of

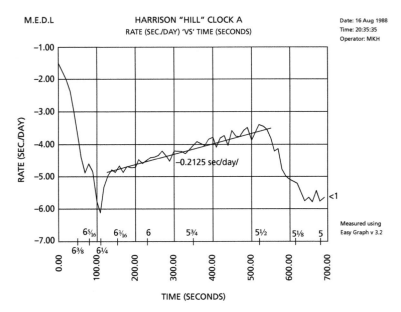

Fig. 9.8 The Hill Test carried out on Clock A with the remontoire locked. © Mervyn Hobden.

Date: 01/26/11 Time: 14:23:57   Date Points 1 thru 21768 of 21768   Tau=2.1000000e+02   File: AVAR Norwich.005

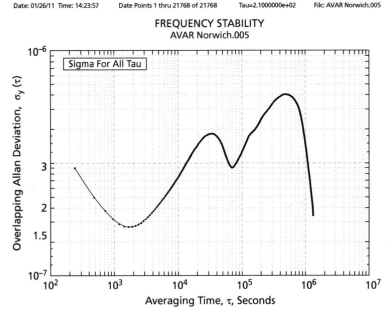

Fig. 9.9 Allan variance measurement of Clock A at Norwich, showing the very good short-term stability over a period of 26 minutes. It also shows the overcompensation for air density against pressure as the sampling increases. © Charles Frodsham & Co.

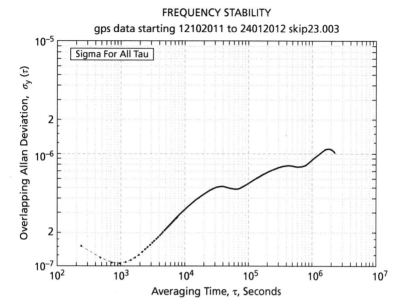

Date: 02/27/12   Time: 17:36:12   Date Points 1 thru 37693 of 37639   Tau=2.4000000e+02   File: gps data starting 12102011to 2

Fig. 9.10 Allan variance measurement of Clock B shows improved short-term stability and undercompensation for air density. © Charles Frodsham & Co.

the characteristic very clearly—and as we later realised, the slope was too steep. As the arc falls with increased pressure and therefore higher density we require that the slope should be positive, but the slope of 0.2125 s/day/0.125° was much greater than that required by the calculation of the flotation error on Martin's pendulum. This result was confirmed by the tests, using the Allan Two-Sample variance carried out by Martin Dorsch of Charles Frodsham & Co. on Clock A at Norwich (Figs. 9.9 and 9.10), which showed that Clock A was over-compensated for the air density flotation error caused by a shift in barometric pressure.

The two plots in Figs. 9.11 and 9.12 show the results of hill tests carried out in November 1986 to confirm the effect of weighting the composers on the grasshopper escapement. The composers soak up the energy released when the pallats interchange, returning the pallats to their rest position. As Harrison said, they have something to do in the matter. Measurements were made with a Beckman averaging frequency counter—the horizontal scale is in setups of the remontoire spring—increasing driving force and the vertical scale is microseconds per

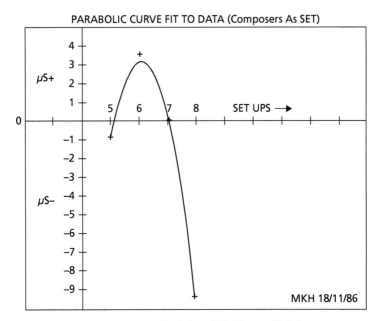

Fig. 9.11 Composers, as set on Clock A in 1986. Mervyn Hobden.

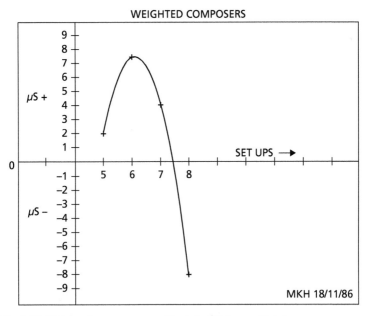

Fig. 9.12 Weighted composers on Clock A. © Mervyn Hobden.

second. A change of 11.57 μs/s is equivalent to 1 s/day change in rate. As can be seen, weighting the composers has changed the magnitude of the hill—from 0.27 to 0.648 s/day. But it has also changed the form of the hill to be almost perfectly parabolic, as the curve fitting shows and that the difference in period output is almost exactly as the difference in (arc) 2 with changing input force. At the running arc, we have a period that is slightly longer than that at the peak of the hill, as required, but in the second plot, because the peak has moved to the right, at the running arc the clock is less overcompensated. The final setting of Clock B was close to this weighted composer plot.

It is necessary to observe changes in the clock period as they occur with changing temperature and pressure. Harrison used the change in phase between the two pendulums of two clocks being tested under different conditions and he tells us that it is possible to compare a change of '1/40 of a second' over a relatively short period. This use of two clocks facilitated adjustment—this is probably the first example of the 'two-oscillator method' now commonly used to facilitate the development of highly stable sources.

A further test carried out on Clock A was to confirm the transient response to a large perturbation (Fig. 9.13)—in the above case from 5.375° to 6.125° − 0.75° increase in semi-arc. This is a large change compared with the pendulum Q, yet the worst-case deviation is less than 0.6 s/day. The first-order transient time is also much shorter than that predicted from the pendulum Q alone. This demonstrates the large energy throughput necessary to allow the clock to adapt quickly to rapidly changing conditions. Hence, the loaded Q must be low and was measured to be 1740 on the basis of energy consumption of the pendulum and escapement unit on Burgess's Clock A.

By reducing the amplitude-dependent nonlinearity or circular deviation, linearizing the grasshopper escapement, and understanding that

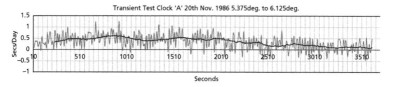

Fig. 9.13 Transient test on Clock A in Martin Burgess's workshop © Mervyn Hobden.

high stored energy (great celerity of motion), reduced external perturbation, Harrison achieved a high degree of dynamic stability. The
fact that amplitude-dependent nonlinearity inevitably leads to high
levels of $1/f$ noise in the oscillator's domain of operation was first con-
clusively proved by David Harrison's PhD thesis of 1988 at Leeds
University. John Harrison was the first person to discover this on his
third timekeeper and it led to changes in the balance spring on his longitude watch H4.

As well as parametric change, it is also necessary to consider the
effect of changing drive force (impulse). Obviously, if the drive force
varies, so will the work done on the pendulum, and there will be a
corresponding change in period. Harrison said that it is easily observed

> that the angle of swing under the action of the grasshopper escapement
> is not in a straight line with respect to the suspension point but at an
> angle set by ... means of the weight of the wires (Harrison, 1763:26)

of the compound pendulum. This can be readily seen in one of Harrison's
early grasshopper regulators with its gridiron pendulum.

Harrison (1763:27) states that 'a greater abatement of force will
attend' so that even if the driving force increases, the period remains
the same. This is analysed by Poynting and Thomson (1920) and in the
case of a pendulum leads to a change in the force acting on the top of
the pendulum rod (Fig. 9.14). A change in driving force is in effect a
change in the forcing angular frequency ω_f acting on the pendulum
with its natural angular frequency ω_n. As demonstrated earlier, there is
a marked phase change in the force acting on the pendulum support as
ω_f passes through the pendulum's resonant frequency.

From an effective forcing period, longer than that of the pendulum
to one shorter than the period of the pendulum, the phase change is
180°—as is found from simple forced resonance theory. When set for a
given exact match to the pendulum's frequency and a constant force,
the 'hastening' effect of the recoil is close to zero due to the 2:3 impulse.
This is no longer correct if the mean force increases and the amplitude
changes and with it the total extent of the recoil arc. This now leads to
a change in the tilt of the pendulum and with it the necessary change
in the recoil force so that the overall period of the system does not
change. This is perhaps the most subtle and insightful of Harrison's
compensations.

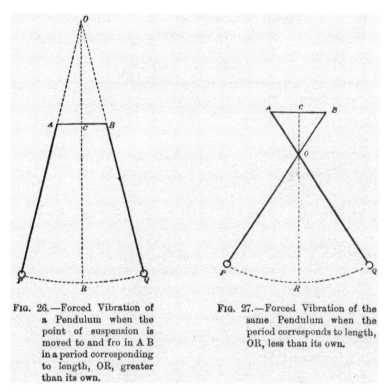

FIG. 26.—Forced Vibration of
a Pendulum when the
point of suspension is
moved to and fro in A B
in a period corresponding
to length, OR, greater
than its own.

FIG. 27.—Forced Vibration of the
same Pendulum when the
period corresponds to length,
OR, less than its own.

Fig. 9.14 Change in driving frequency effect on a compound pendulum around its natural frequency of oscillation (Poynting and Thompson, 1920).

Combined with the other parametric compensations, it leads to an oscillator that is truly adiabatically invariant under perturbation. (An adiabatic oscillator is one where slow changes in any parameter are assumed to have a negligible effect, if the change is very slow compared with the period of oscillation. In physics it is viewed as approximation to a conservative system.) It is demanded that the linear oscillator which is the basis of quantum mechanics be adiabatically invariant; however, the proof offered by Ehrenfest in 1914 is not physically rigorous (Born, 1948:108–112). As Harrison demonstrates, only a nonlinear oscillator is capable of such performance.

Unlike later analysts, who attempted to fit experimental results into a preconceived algebraic model, Harrison analysed what was there and from that was able to extract the 'characteristic function' of the

oscillator's operation. The above is a brief introduction to the depth
of Harrison's understanding. Harrison (1775:59) loudly proclaimed his
contempt for the 'Professors of Arts or Sciences at Cambridge and
Oxford, as from their high Algebra'

He also complained of their low level of understanding of mechanics: 'They can indeed tell us of what will be the Result of the Motion or
Motions of two Marbles (such as Boys play withal) rapping or impin-
ging one against the other' (Harrison, 1775:52). There is some justice in
his complaints, William Ludlam, a member of the Board of Longitude,
acknowledged that Harrison's design was 'the result of sound reasoning, though perhaps not dressed out in all the formality of theorems
and corollaries' (Ludlam, 1769:135).

After the death of Newton, the heirs to his legacy, the 'Gentlemen
well skilled in Mechanics' of whom Harrison so bitterly complained,
carried out little original work in mechanics and it is not until the 1780s
that a revival occurred in the development of Newton's fluxional calculus (Guicciardini, 1989), as shown by George Atwood's paper where
he analyses the complex operation of Mudge's marine timekeeper's
escapement (Atwood, 1795). This revival was far too late to be significant as it was rapidly submerged beneath the steady state 'certainties' of
Lagrange and Laplace.

Perhaps the best epithet to Harrison's life was recorded by Rupert
Gould (2013:70):

> But his truest memorial is to be found in the hearts of those who know
> and appreciate the pioneer work of the man who lies there
> '... still loftier than the world suspects, Living and dying...'

Acknowledgements

The work of the Harrison Research Group was only possible because a
group of people pooled their knowledge and expertise to find out the
truth about John Harrison's legacy. Martin Burgess's selfless commit-
ment to the project to build two regulators provided a focus around
which we could all move. We dedicated what time we could spare from
busy careers and lifestyles, but Martin and Eleanor provided the retreat
to which we could always recourse to discuss and argue out our mutual
fascination.

References

Acasu, Z. (1954). Vander Pol's equation: An analytic method of general solution. *The Wireless Engineer*, August, 198–203.

Airy, G. B. (1827). On disturbances of pendulums and balances and on the theory of escapements. *Transactions of the Cambridge Philosophical Society*, 3, Part I, 105–128.

Atwood, G. (1795). Investigations, founded on the theory of motion, for determining the times of vibration of watch balances. *Philosophical Transactions of the Royal Society*. London

Born, M. (1948). *Atomic Physics*, 4th edition. London: Blackie & Sons

Burgess, M. (1996). The scandalous neglect of Harrison's regulator science. In: Andrewes, W., ed., *The Quest for Longitude: The Proceedings of the Longitude Symposium, Harvard University, Cambridge Massachusetts, November 4–6, 1993*, pp. 256–78. Harvard, Collection of Historical Scientific Instruments.

Gould, R. T. (2013). *The Marine Chronometer, Its History and Development*. Woodbridge: Antique Collector's Club.

Groszkowski, J. (1933). The interdependence of frequency variation and harmonic content and the problem of constant frequency oscillators. *Proceedings of Institute of Radio Engineers*, 21(7), 958–979.

Guicciardini, N. (1989). *The Development of the Newtonian Calculus in Britain, 1700–1800*. Cambridge: Cambridge University Press.

Harrison, J. (1730). *The 1730 manuscript*. Refer to Bromley, J. (1977) *The Clockmakers' Library: the catalogue of the books and manuscripts in the library of the Worshipful Company of Clockmakers*. London, Sotheby Parke Bernet Publications, P.108. MS977.

Harrison, J. (1763). *An Explanation of My Watch*. Available at https://ahsoc.contentfiles.net/media/assets/file/Explaining_My_Watch.pdf. Accessed 4 December 2018.

Harrison, J. (1775). *A Description Concerning Such Mechanism*. London, Jones, T. Available at https://ahsoc.contentfiles.net/media/assets/file/Concerning_Such_Mechanism.pdf. Accessed 12 December 2018.

Hobden, M. (1982). John Harrison, Balthazar van der Pol and the non-linear oscillator, Part 4. *Horological Journal*, April, pp. 15–18.

Hobden, M. (2011). As 3 is to 2. *Horological Science Newsletter,* Issue 2011–5, pp. 2–13.

Hobden, M. K. (2015). Dominion and dynamic stability. *Horological Science News*, 2015–4, p. 9.

Jeans, J. H. (1925). *The Dynamical Theory of Gases*. Cambridge: Cambridge University Press.

Le Corbellier, P. H. (1960). Two stroke oscillators. *IRE Transactions on Circuit Theory*, December, 387–398.

Ludlam, W. (1769). *Astronomical Observations Made In St. John's College, Cambridge, in the Years 1767 and 1768: With an Account of Several Astronomical Instruments*. Cambridge: Archdeacon, J.

Newton, I. (2010). *The Principia*, 3rd edition. Illinois: Snowball Publishing.

Poynting, J. H., and Thomson, J. J. (1920). *Text-Book of Physics: Sound*. London: Charles Griffin.

Quill, H. (1966). *John Harrison, The Man Who Found Longitude*. London: Baker.

Robertson, D. (1928–32). The theory of pendulums and escapements. *Horological Journal*, September 1928 to February 1932.

Smith, M. K. (1964). Precision measurement of period vs amplitude for a pendulum. *American Journal of Physics*, 32(8).

Strutt, J. W. (1883). On maintained vibrations. *Philosophical Magazine*, 15, 229–231.

Taylor, E. G. R. (1954). *The Mathematical Practitioners of Stuart and Tudor England*. Cambridge: Cambridge University Press.

Van der Pol, B. (1934). The non-linear theory of electronic oscillators. *Proceedings of the Institute of Radio Engineers*, 22(9), 1051–1083.

10

Analysis of the Mechanisms for Compensation in Clock B

David Harrison

Introduction

John Harrison described a radically new approach to improving the timekeeping accuracy of clocks in his (1775) book, *A Description Concerning Such Mechanism as Will Afford a Nice, or True Mensuration of Time*. In this, he set out the key elements of his approach, which included the use of the following:

- A large running arc to increase pendulum velocity and the effect of air resistance on its motion;
- A grasshopper escapement to remove the need for oil and provide an increasing torque during the pendulum swing to compensate for changes in air density;
- Weighted composers to tune the effect of the grasshopper escapement on the pendulum rate;
- A thin stiff suspension spring and circular cheeks to modify the escapement torque applied to the pendulum and partially compensate for the pendulum's circular error;
- A modest bob mass to prevent straining the suspension spring; and
- A pendulum length that reduces slightly when warm.

Harrison described how the use of these features when used in combination would provide significantly better timekeeping performance than more conventional designs using a high bob mass and a small running arc. A version of one of Harrison's pendulum clock designs, incorporating the features outlined above, has provided physical

Harrison, D., *Analysis of the Mechanisms for Compensation in Clock B* In: *Harrison Decoded: Towards a Perfect Pendulum Clock*. Edited by Rory McEvoy and Jonathan Betts, Oxford University Press (2020). © Oxford University Press. DOI: 10.1093/oso/9780198816812.003.0010 evidence supporting these claims; with this clock now consistently providing an accuracy of better than a second in a hundred days (Betts, 2016).

This paper provides a more complete mathematical analysis of how Harrison achieved this improved timekeeping performance. The three key elements of his approach are explained:

1. Firstly, cancelling, for all practical purposes, the effects of changes in air density and viscosity on the pendulum frequency. Harrison's approach enables this cancellation to be achieved over a much wider range of changes in air density and viscosity than more conventional pendulum clock designs. This enabled Harrison to achieve improved longer-term timekeeping performance, because over longer time periods it is more likely that more significant changes in air density and viscosity will occur with more extreme changes in air: temperature, pressure and humidity. These larger changes would affect the longer-term timekeeping performance of conventional clocks but have no significant effect on clocks made using Harrison's design principles.

2. Secondly, cancelling for all practical purposes, the effects of changes in escapement torque provided by the remontoir spring on the pendulum rate;

3. Thirdly, the use of a low pendulum quality factor Q by using a modest bob mass and a large running arc. This allows the running arc to return more quickly to its free running level following perturbations by external forces such as the vibrations from passing traffic and draughts. This suppresses the random variations in the pendulum running arc and their cumulative longer-term effects on the timekeeping performance of the clock.

Harrison's design approach used the inherent nonlinearities in the motion of a physical pendulum to his advantage as opposed to trying to minimise them. To achieve this, he used a pendulum with a large running arc. This increases the nonlinear effects of the escapement and the circular errors on the pendulum frequency with changes in running arc. As described later, Harrison tailored these nonlinear effects to his needs using the grasshopper escapement geometry and a thin stiff suspension spring in combination with circular cheeks. Their contributions to the pendulum frequency with changes in running arc (resulting from changes in air density) were used to cancel for all practical purposes other direct changes in frequency caused by changes in the pendulum buoyancy and the mass of air it displaces with air density. He also provided an additional compensating mechanism for changes in air viscosity using a pendulum rod length that reduces slightly with temperature.

Harrison also used the grasshopper escapement geometry and a large running and escaping arc to cancel for all practical purposes the effect of changes in escapement torque on the pendulum rate.

Harrison's approach represented a significant departure from the conventional clock design approach of trying to remove the escapement and circular error nonlinearities as far as possible, using a small running arc and a high bob mass. A lack of understanding of Harrison's nonlinear approach when his (1775) book was published (a lack which still largely remains today) led Harrison to remark on page 4 of that work:

> sometimes some Men, as being quite ignorant in what I am here about to shew or speak of, and as when they are about to do something very extraordinary as they imagine, do render the Matter as still worse than so, yea even by far.

Changes in air density and viscosity on the pendulum rate

There are two mechanisms for how changes in air density and viscosity affect the frequency of pendulum oscillations:

1. *Direct changes to the frequency of the pendulum:* Here changes in the air density due to variations in temperature, pressure and humidity change the buoyancy of pendulum bob and the mass of air it displaces (Nelson & Olsson, 1986). Both cause a direct reduction in the frequency of the pendulum oscillations when the density of the air increases. In contrast, changes in air viscosity with temperature do not cause these changes.
   These direct changes are not dependent on the running arc of the pendulum.
2. *Indirect changes to the frequency of the pendulum:* Here changes in the density and viscosity of the air cause a change in its resistance to the motion of the bob. These cause changes to the pendulum running arc, which are converted by the nonlinear escapement and pendulum circular errors into a change in the pendulum frequency. These indirect changes are dependent on the pendulum's running arc.

Harrison's approach involves making the direct and indirect changes in pendulum frequency with changing air density and viscosity cancel over the full potential range of changes in temperature, pressure and humidity, so they have no meaningful effect on the pendulum's time-keeping performance.

To achieve this, Harrison used the following:

- A large pendulum running arc to increase the required escapement torque. This increases the escapement's indirect effect on the pendulum frequency;
- A grasshopper escapement torque that increases during the pendulum swing, where the ratio of the torques before and after the pallets interchange is approximately 3:2. This establishes the grasshopper's effect on the pendulum frequency;
- Weighted composers to modify the increase in escapement torque before and after the escaping arcs are reached. These provide a practical means of tuning the effect of the grasshopper escapement on the pendulum frequency;
- A thin stiff suspension spring of specific thickness (and hence stiffness) and circular cheeks. These partially compensate for the circular error; and increase the required escapement force, reducing the effect of random changes in this force; and
- A pendulum rod length that reduces slightly when warm. This provides the required direct change in pendulum rate with changes in air viscosity caused by changes in temperature. Without this additional direct change in the pendulum frequency, it would be impossible to cancel the indirect changes in pendulum rate caused by changes in its running arc due to changes in viscosity.

As we show later, more conventional pendulum clock designs are also able to achieve some cancellation of the direct and indirect changes of the pendulum frequency with air density, by using the correct ratio between their escaping and running arcs. However, unlike Harrison's approach this cancellation does not work over the full potential range of changes in air density and viscosity with temperature, pressure and humidity. This means that unlike Harrison's clock design, the time-keeping performance of conventional clocks is affected when larger changes in air density and viscosity occur.

In addition, Harrison's use of low pendulum Q reduces the random variations in the pendulum's running arc caused by external perturbing forces, and hence the changes they cause to the pendulum frequency.

The indirect effect of the grasshopper escapement on the pendulum rate with changes in running arc

In this section we show for Harrison's pendulum clock design how changes in its running arc, due to changes in air resistance, change its frequency. Because Harrison's grasshopper escapement does not require oil, changes in running arc are mainly caused by changes in air resistance as opposed to changes in the clock's mechanical resistance. Changes in air resistance are caused by changes in air density and viscosity. Air density changes with variation in temperature, pressure and humidity, whilst air viscosity changes only with temperature.

Figure 10.1 shows a simplified illustration of Harrison's grasshopper escapement. The escapement torque increases with pendulum arc before the escaping arc is reached, due to an increase in the torque arm radius R. Harrison recommended that the ratio of the mean torque applied before and after the pallets interchange (at the escaping arc) should be ~3:2. Harrison (1775:25–26) referred to this as follows:

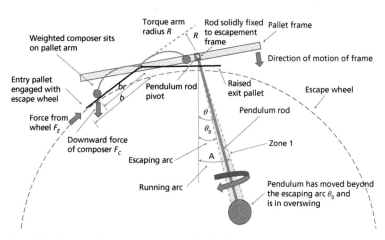

Fig. 10.1 The grasshopper escapement (with the pendulum in supplementary arc). © David Harrison

But as here, to come a little nearer in this material Point, let, as I order the Matter, the Force (from the Wheel) upon the Pendulum, as just before the interchanging of the Pallats, to be as by or from them the said Pallats supposed or taken as 3, then, as just after their interchanging (and the Force to contrary Direction, it must but be about as 2, that is, it must be so ordered, (as may hereafter be observed by the Drawing) viz. as that it be so by the taking, or supposing for the Purpose, a Mean betwixt the Actions of each Pallat, and withal, as farther to the Purpose, that, as in the little recoiling of the Wheel, to become less and less to the Extremity of each Vibration, but as whence, or as still on Course, the greater at any Time the whole Vibration may be, of more Efficacy the same small Force (and still as it were in the little recoiling) must from its Quantity or Duration prove, and that in such small Measure as required;

A geometric analysis of Harrison's drawing of the grasshopper escapement (Heskin, 2011) reveals that it does provide an increase in torque with pendulum arc θ given by

$$\tau(\theta) = \mu(1 + \alpha\theta),   \tag{10.1}$$

where μ is the constant level of escapement torque switched in direction every half-cycle of the pendulum swing at the escaping angles, $\pm\theta_0$. From the drawing, the rate of increase in torque α with pendulum arc is found to be approximately 2.8, averaged over the two halves of the pendulum swing. This provides, as Harrison recommended, an average ratio between the torques applied before and after the pallets interchange of ~3: 2.

In practice, the single pivot grasshopper pallet geometry provides a different rate of increase in escapement torque with pendulum arc during its forward and backward swings. This does not change their combined effect on the pendulum rate and is therefore not an essential part of Harrison's approach. Harrison (1775:24) said:

> For it may be notified, as just here hinted, that the Actions of each Pallat are not equally the same upon the Pendulum, but not so, as to be easily perceived to be otherwise, viz. as in the looking at the Clock, or Seconds in Motion, although in this most highly material Circumstance, or Construction of the Pallats, pretty much different, but still, not so to be taken (viz. as with respect to the essential Point here in Hand) as to be any the least worse for the same, but as that the Action of one with that of the other, are quite right

The torque applied by the grasshopper when the pendulum is travelling from $-A$ to A is shown in Fig. 10.2.

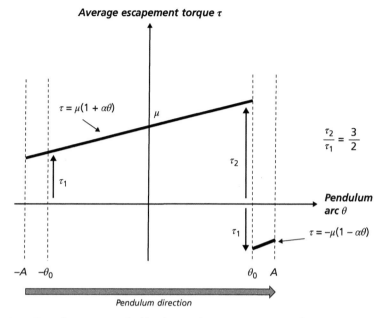

Fig. 10.2 The torque applied by the grasshopper escapement. © David Harrison

The dependency caused by the grasshopper escapement torque shown in Fig. 10.2 between the change in pendulum frequency $\Delta\omega$ and running arc A is, using Eq. (10.B1) in this chapter's Annex B, given by

$$\Delta\omega = \frac{2\mu}{\omega_0 \pi A}\sqrt{1-\left(\frac{\theta_0}{A}\right)^2} + \frac{\mu\alpha\theta_0}{\omega_0 \pi A}\sqrt{1-\left(\frac{\theta_0}{A}\right)^2}$$
$$-\frac{\mu\alpha \sin^{-1}\left(\frac{\theta_0}{A}\right)}{\omega_0 \pi}$$

(10.2)

Here, $\Delta\omega$ is the change in pendulum frequency with running arc A, and θ_0 is the escaping angle when the pallets interchange, reversing the escapement torque. Harrison said that the pendulum running arc should be large. This increases the required sustaining escapement torque μ and hence from Eq. (10.2) the indirect effect of the escapement error on the pendulum frequency with changes in running arc.

So far we have not considered the effect of the weighted composers shown in Figs. 10.1 and 10.3. These cause a change in the gradient of the grasshopper escapement torque before and after the escaping arc is reached. This is because, as illustrated in Figs. 10.1 and 10.3, they change their resting positions before and after the escaping arc is reached.

In Fig. 10.1, when the pendulum is in its supplementary arc in zone 1, the weighted composer applies a downward force to the entry pallet arm. Whilst in Fig. 10.3, when the pendulum is moving within the escaping arcs in zone 2, the weighted composer applies a downward force to the pallet frame. An analysis of the forces applied to the pendulum by the composers in these distinct positions is provided in this chapter's Annex A. From this, the change in position of the composers during the pendulum swing is found to increase the slope of the applied escapement torque with pendulum arc in zone 1, and to reduce it in zone 2.

When the pendulum is travelling between $-A$ and θ_0, the grasshopper torque is given by

$$\tau = \mu(1+\alpha_1\theta), \tag{10.3}$$

where α_1 is the increased torque gradient in zone 1 caused by the weighted composers. Similarly, when the pendulum is travelling between $-\theta_0$ and θ_0, the grasshopper torque is given by

$$\tau = \mu(1+\alpha_2\theta), \tag{10.4}$$

where α_2 is the reduced torque gradient in zone 2 caused by the weighted composers.

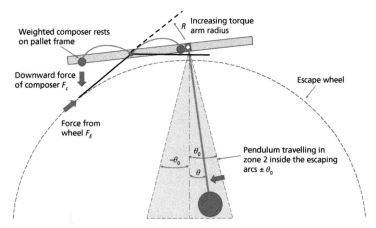

Fig. 10.3 The grasshopper escapement with the pendulum inside the escaping arc. © David Harrison

The dependency caused by these torques between the pendulum frequency and its running arc is found using Eq. (10.B1) in Annex B to be

$$\Delta\omega = \frac{2\mu}{\omega_0 \pi A}\sqrt{1-\left(\frac{\theta_0}{A}\right)^2} + \frac{\mu\theta_0\alpha_2}{\omega_0\pi A}\sqrt{1-\left(\frac{\theta_0}{A}\right)^2}$$
$$-\frac{\mu\theta_0\alpha_2}{\omega_0\pi}\sin^{-1}\left(\frac{\theta_0}{A}\right). \tag{10.5}$$

Here $\Delta\omega$ is the change in the frequency of the pendulum from its natural frequency ω_0 caused by the applied escapement torque. A is the running arc of the pendulum, and θ_0 is the escaping angle when the pallets interchange.

From Eq. (10.5) by adjusting the value of α_2, by changing the mass and position of the composers, the change in pendulum frequency with running arc caused by the grasshopper escapement can be finely adjusted.

Harrison (1775:50) referred to using the composers to adjust the clock's timekeeping performance in the following terms:

> There being to be concerned in that Proceeding, four different Things, and wherein two of which (as touching the Point) do as it were pretty much conspire to, or in the same Purpose, viz. the Composers of the Pallats to relative Rest, and the correspondent Curvature thereto of the Cycloid Cheeks. I say, these two Things may only as almost be taken as one, viz. in their joint Effects, for so far as belongs to this Matter, but not quite so.

Equation (10.5) is different from that derived previously for Harrison's grasshopper escapement primarily because its torque was taken to increase as opposed to decrease when the pendulum moves beyond the escaping arc and is in supplementary arc (Woodward, 1997; Haine and Millington, 2016).

The indirect effect of the suspension spring and circular cheeks on the pendulum rate with changes in running arc

Another important aspect of Harrison's pendulum clock design was his use of a thin stiff suspension spring of a specific thickness (and hence stiffness) with circular cheeks, shown in Fig. 10.4.

These have the following effects on the pendulum:

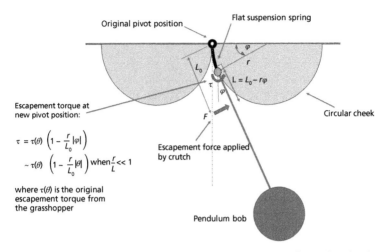

Fig. 10.4 Pendulum with a thin stiff suspension spring and circular cheeks.
© David Harrison

1. The spring wrapping around the cheeks with increasing arc reduces the effective length of the pendulum and its potential energy. This is because the *effective* pendulum pivot point moves down and around the cheeks with increasing arc. These effects do not fully cancel the circular error and add a linear increase to the pendulum frequency with running arc;

2. The physical stiffness of the spring applies an torque opposing the pendulum motion with increasing displacement arc. This partially reduces the circular error;

3. The spring and cheeks reduce the grasshopper escapement torque applied to the pendulum with displacement arc. This change is caused by the reducing torque arm radius L shown in Fig. 10.4 with increasing arc. This change increases the required force from the escapement spring, reducing the effect of small unwanted changes in this force on the motion of the pendulum.

We consider these in more detail in the next section.

Modification of the pendulum circular error by the spring and circular cheeks

The use of the large pendulum running arc used in Harrison's clock design increases the pendulum circular error and the dependency of the pendulum rate on its running arc A, given by

$$\Delta\omega = -\omega_0 \frac{A^2}{16}. \tag{10.6}$$

Harrison used the thin suspension spring and circular cheeks shown in Fig. 10.4 to control the circular error. These have two effects on the changes in pendulum frequency with running arc:

1. The spring wrapping around the cheeks reduces the effective length of the pendulum and its potential energy with increasing arc; and
2. The bending of the stiff thin suspension spring provides a torque opposing the pendulum motion with increasing displacement arc.

The first of these has been shown to cause a linear change in pendulum frequency with running arc (Mazaheri et al., 2012) given by

$$\Delta\omega \sim \omega_0 \frac{2\beta A}{3\pi}. \tag{10.7}$$

Here, $\Delta\omega$ is the change in pendulum frequency with running arc A, and β is the ratio of the radius of the circular cheeks to the length of the pendulum rod l.

The second is shown in annex C to cause a change in pendulum frequency with running arc given by

$$\Delta\omega \sim \frac{k}{\omega_0}\left(\frac{A^2}{16} - \frac{1}{2}\right). \tag{10.8}$$

Here, k is a constant relating to the spring stiffness.

Adding Eqs. (10.5), (10.6), (10.7) and (10.8) we find the total indirect change in pendulum frequency with running arc due to the grasshopper escapement, spring and circular cheeks to be

$$\Delta\omega = \frac{2\mu}{\omega_0 \pi A}\sqrt{1-\left(\frac{\theta_0}{A}\right)^2} + \frac{\mu\theta_0\alpha_2}{\omega_0\pi A}\sqrt{1-\left(\frac{\theta_0}{A}\right)^2}$$
$$-\frac{\mu\theta_0\alpha_2}{\omega_0\pi}\sin^{-1}\left(\frac{\theta_0}{A}\right) - \frac{A^2\omega_0}{16}\left(1-\frac{k}{\omega_0^2}\right) + \frac{2\beta A\omega_0}{3\pi} - \frac{k}{2\omega_0}. \tag{10.9}$$

This shows that by adjusting the escapement torque μ, the torque gradient provided by the grasshopper escapement α_2, using its geometry and the composers; the cheek radius ratio β; and the stiffness of the spring k (by adjusting its modulus of elastisty, thickness and length), the dependency of the pendulum frequency on its running arc can be controlled.

Modification of the escapement force by the spring and circular cheeks

From Fig. 10.4, the torque arm radius L reduces with increasing arc and total escapement torque applied to the pendulum with circular cheeks and thin suspension spring is given by

$$\tau = \tau(\theta)(1 - \frac{r}{L_0}|\theta|)$$
$$= \tau(\theta)(1 - \alpha_3|\theta|).$$

The modulus sign is used around the displacement angle θ because the torque arm radius reduces for both positive and negative displacement angles. $\tau(\theta)$ is the original grasshopper escapement torque given in Eq. (10.1), and α_3 is the ratio of the radius of the cheeks to the torque arm radius L_0 at the pendulum zero displacement angle. Using Eq. (10.1) the resulting escapement torque applied to the pendulum is given by

$$\tau = \mu\left(1 + \alpha\theta - \alpha_3|\theta| - \alpha\alpha_3|\theta|\theta\right). \tag{10.10}$$

Comparing Eqs. (10.10) and (10.1) we can see that two additional terms have been added to the original grasshopper torque: $-\alpha_3|\theta|, -\alpha\alpha_3|\theta|\theta$. The first, $-\mu\alpha_3|\theta|$, reduces the energy provided by escapement remontoir spring to the pendulum by $\mu\alpha_3 A^2/2\pi\omega$. This increases the required force from the escapement spring, reducing the effect of small variations in this force on the pendulum.

Assuming that air resistance increases with velocity, the running arc of the pendulum is given (Hoyng, 2014) by

$$A_0 = \sqrt{\frac{2\mu\theta_0}{\pi\omega_0\epsilon_0}}, \tag{10.11}$$

where the air's resistance ϵ_0 is related to the pendulum quality factor Q by $\epsilon_0 = \omega_0/2Q$ and μ is the torque provided by the escapement.

Using Eq. (10.11) changes in running arc caused by changes in air resistance $\Delta\epsilon_0$ due to changes in air density and viscosity can be calculated using the equation

$$A_0 = \sqrt{\frac{2\mu\theta_0}{\pi\omega_0(\epsilon_0 + \Delta\epsilon_0)}}. \tag{10.12}$$

Inserting the pendulum running arc A given by Eq. (10.11) into Eq. (10.8) we can calculate for Harrison's clock the total change in pendulum frequency caused by the grasshopper escapement and the spring and cheeks with changes in air resistance, $\Delta\epsilon_0$. We will return to this later.

Harrison (1775:46) referred to the increase in pendulum frequency and reductions in running arc caused by the spring and circular cheeks in the following terms:

> Now the Nature of such a Matter, or Cycloid to the Purpose, (and as consequently withal for preserving the Spring) must be as in some Measure reverse to what is demonstrated by Mr Huygens, &c. that is, it must be so as to occasion little Vibrations of the Pendulum, viz. all such as are less (and unregarded) than so as to let, or such as will let the Pallats interchange, to be still sooner performed, than what they would as otherwise be without it;

The direct changes in pendulum frequency due to changes in air density

As highlighted previously (Nelson and Olsson, 1986; Belleville, 2007), changes in air density also cause direct changes to the pendulum frequency. These occur because changes in air density change the pendulum buoyancy and the mass of air its motion displaces. The pendulum frequency with these direct changes is given by

$$\omega = \omega_0 \frac{\sqrt{1 - k_1 \dfrac{\rho_a}{\rho_b}}}{\sqrt{1 + \dfrac{\rho_a}{\rho_b}}}, \tag{10.13}$$

where ρ_a is the density of the air and ρ_b is the density of the bob. k_1 is a constant equal to 1 for a simple pendulum. For a compound pendulum it is less than 1 because the centre of buoyancy is located above centre of the bob due to the mass of air displaced by the rod. For

Harrison's pendulum $k_1 \sim\sim 0.75$. From Eq. (10.12) these changes are not dependent on the running arc of pendulum.

Changes in air viscosity do not cause direct changes in pendulum frequency, which we will return to later.

Cancelling the effect of changes in air density on the pendulum rate

We now show how Harrison arranged for the direct and indirect effects of changes in pendulum frequency caused by changes in air density to cancel. First, we use Eq. (10.11) to calculate the running arc after a change in air resistance $\Delta\epsilon_0$ due to a change in air density. Changes in air density occur due to changes in temperature, pressure and humidity, and can vary by as much as $\pm10\%$ between a cool humid day and a hot dry day (Belleville, 2007).

Using Eq. (10.8) we then calculate the indirect changes in pendulum frequency caused by the grasshopper escapement with the suspension spring and circular cheeks. Using Eq (10.13) we also calculate the corresponding direct changes in pendulum frequency with changes in air density due to changes in the pendulum buoyancy and the mass of air displaced by the motion of the bob. By summing the indirect and direct changes we calculate the total change in pendulum frequency with air density shown in Fig. 10.5.

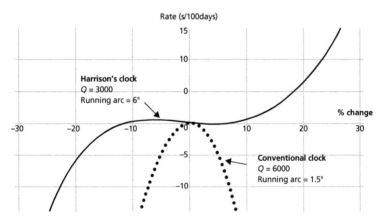

Fig. 10.5 Changes in clock rate with changes in air density for Harrison's and a more conventional high-Q clock. © David Harrison

The curves shown in Fig. 10.5 were produced using an interactive graphing tool. A combination of a grasshopper escapement torque gradient $\alpha = 2.8$ (corresponding to the a 3:2 torque ratio specified by Harrison), and a cheek radius of 6.35 cm, and reduction of the circular error by 10% was found to cancel the direct and indirect effects of changes in air density on the pendulum rate over the full potential range of change in air density of $\pm 10\%$. Different potential combinations of these parameters can provide similar compensation results. The compensation is primarily controlled by the radius of the cheeks and the grasshopper torque gradient.

Harrison (1775:47) referred to the dependency of his nonlinear compensation approach on both the escapement geometry and the cheeks as

'but however, as each Cheek, with regard to the Property I have shewn of the Pallats, or as a Tenor to their Result, must be the Arch of a Circle'.

Figure 10.5 also compares the rate performance achieved by a more conventional high-Q, small running arc clock design. This clock has a Q of 6000 and a running arc of 1.5°. It uses a constant torque recoil escapement, where the escaping arc has been adjusted to minimise the effect of changes in air density on the pendulum rate at the running arc of the pendulum.

Figure 10.5 shows the clear level of improvement provided by Harrison's approach. Whilst Harrison's clock delivers virtually no dependency between the pendulum rate and changes in air density over the full potential range of changes in air density of $\pm 10\%$, the conventional clock design causes a much more significant hill-shaped dependency. Figure 10.5 shows that whilst Harrison's clock is able to achieve better than a second in a hundred days' accuracy, even with large changes in air density, the conventional higher Q clock design would be unable to do so.

Cancelling the effect of changes in air viscosity on the pendulum rate

Without additional steps being taken, the approach described in the previous section for cancelling the indirect and direct effects of air density on the pendulum rate will not work for changes in air viscosity, as explained by Hobden (Chapter 9). This is because changes in air viscosity do not cause changes in changes in buoyancy and frequency, which are needed to cancel the indirect changes caused by changing air resistance with viscosity.

Air viscosity changes with temperature and is given by (Schlichting and Gersten, 2003)

$$v = v_0 \left(\frac{T}{T_0} \right)^{0.7} , \qquad (10.14)$$

where T_0 is 273° Kelvin, and T is the temperature of the air.

Using Eqs. (10.9), (10.12) and (10.14) the pendulum rate is found to reduce with increasing temperature and hence viscosity, when Harrison's pendulum clock has been adjusted as described in the previous section to cancel the effects of changes in air density on its rate. To provide the required direct compensating mechanism for changes in viscosity, Harrison used a pendulum rod length that reduces slightly with temperature. This approach is seen in Fig. 10.6 to significantly reduce the dependency of the pendulum rate on temperature.

Harrison (1775:25) explained the need for this additional compensation mechanism:

> but not to be the whole Concern in that Matter; the Pendulum withal requiring to be (viz. as from my Contrivance of its Combination of Brass and Steel Wires) rather, as mathematically speaking, shorter when warm than when cold.

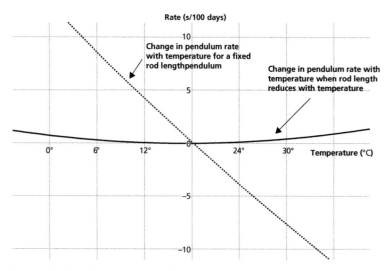

Fig. 10.6 The dependency of Harrison's pendulum clock on changes in viscosity with air temperature. © David Harrison

Minimising the effect of changes in the escapement torque on the pendulum rate

When Harrison's clock is optimised to reduce the effects of changes in air resistance on its rate, a residual hill-shaped dependency is found between changes in the escapement torque and the pendulum rate. The peak of this hill occurs below the running arc of the pendulum. This is consistent with the practical experience of adjusting Harrison's clocks for best air density performance, which occurs when the pendulum rate decreases slightly with increasing escapement torque (Betts, 2016).

Harrison took additional steps to reduce the effect of the residual dependency between the pendulum rate and the escapement torque. These included reducing friction and the need for oil in his grasshopper escapement; using a large running arc and hence large escapement torque so that small changes in the escapement torque have less effect on the pendulum rate; and using a remontoir to provide a more constant driving torque.

From Eq. (10.2) changes in escapement torque also cause direct changes in pendulum frequency. For example, when the force from the escapement remontoir spring increases by $\Delta\mu$ during the pendulum swing, the grasshopper escapement torque becomes

$$\tau(\theta)=\left(\mu_0+\frac{\Delta\mu\theta}{A}+\Delta\mu\right)(1+\alpha_2\theta).$$

The resulting changes in escapement torque are

$$\Delta\tau(\theta)=\Delta\mu+\Delta\mu\theta\alpha_2+\frac{\Delta\mu\theta}{A}+\frac{\Delta\mu\alpha_2\theta^2}{A},$$

which using Eq. (10.B1) cause the following changes in pendulum frequency:

$$\Delta\omega=\frac{2\Delta\mu}{\omega_0\pi A}\sqrt{1-\left(\frac{\theta_0}{A}\right)^2}+\frac{\Delta\mu\theta_0\alpha_2}{\omega_0\pi A}\sqrt{1-\left(\frac{\theta_0}{A}\right)^2}$$

$$-\frac{\Delta\mu\alpha_2}{\omega_0\pi}\sin^{-1}\left(\frac{\theta_0}{A}\right)+\frac{\Delta\mu\theta_0}{\omega_0\pi A^2}\sqrt{1-\left(\frac{\theta_0}{A}\right)^2}-\frac{\Delta\mu\sin^{-1}\left(\frac{\theta_0}{A}\right)}{\omega_0\pi A}$$

$$+\frac{2\Delta\mu\alpha_2\left(\theta_0^2+2A^2\right)}{3\omega_0\pi A^2}\sqrt{1-\left(\frac{\theta_0}{A}\right)^2}.$$

These changes in pendulum frequency cancel when a large escaping arc θ_0 and a grasshopper torque ratio of 3:2 is used. This means that for Harrison's clock, changes in escapement torque have for all practical purposes no direct effect on the pendulum frequency.

A larger pendulum running arc gives Harrison's clocks other additional benefits

The use of a large pendulum running arc and low pendulum bob mass recommended by Harrison also helps reduce the effect of external perturbing forces on the pendulum rate.

A simplified and linearised analysis of the effect of external random forces on the rate of an oscillator found that their effects cancel as time progresses, and hence have no effect on the longer-term timekeeping performance of a clock (Leeson, 1966). However, a more recent and more comprehensive analysis has found that external random forces cause the timing error of the oscillations to accumulate with time (Demir et al., 2000). The variance of the accumulated timing jitter caused by the external perturbations is given by

$$\sigma_{acc}^2 = ct, \tag{10.15}$$

where σ_{acc} is the standard deviation of the accumulated timing jitter of the oscillations, and c is a constant for a given oscillator. For a pendulum oscillator, the constant c is given by

$$c = \left(\frac{rms\ level\ of\ external\ force}{2ml^2 A} \right)^2, \tag{10.16}$$

where l is the length of the pendulum and m is the mass of the bob.

Equation (10.15) shows that the lighter bob mass used by Harrison (to prevent straining and deforming the thin suspension spring) when used in conjunction with a higher running arc A is able to achieve a lower level of accumulated timing jitter performance than a heavier pendulum bob clock with a lower running arc. Using the clock parameters shown in Fig. 10.5, Harrison's clock has approximately half the timing jitter of the higher Q clock.

So far, we have only considered the *direct* effect of external random perturbations on changes to the pendulum rate. In addition, there is an *indirect* contribution to the timing jitter caused by the external perturbations causing random changes to the running arc of the pendulum which are converted by the escapement and circular errors into random changes in pendulum frequency (see Eq. (10.8)). These indirect contributions to the pendulum rate are lower in Harrison's clock design than in more conventional high-Q clocks because of its lower Q pendulum. We explore this in more detail later.

It has been shown (Hoyng, 2014) that the change in pendulum running arc with time is given by

$$\frac{dA^2}{dt} = -2\epsilon_0 A^2 + \frac{4\mu\theta_0}{\pi\omega_0}. \tag{10.17}$$

After a small change in running arc δA caused by an external perturbation, the running arc is given by

$$A = A_0 + \delta A, \tag{10.18}$$

where A_0 is the pendulum's steady running arc.

Substituting Eq. (10.17) into Eq. (10.16) and using Eq. (10.7), we find that

$$\frac{d\delta A}{dt} \sim -\frac{\omega_0}{Q}\delta A. \tag{10.19}$$

The solution to Eq. (10.18) is

$$\delta A(t) = \delta A_0 e^{-\frac{\omega_0}{Q}t}, \tag{10.20}$$

where δA_0 is the initial displacement in running arc by the external perturbation at $t = 0$.

From Eq. (10.19), small changes in running arc δA due to external perturbations decay exponentially with a time constant Q/ω_0. Hence, Harrison's use of a low pendulum Q enables its running arc to return faster to its steady value following perturbations by external forces. This faster return reduces the random variations in the pendulum running arc to a lower level than in more conventional higher-Q clock designs. The increased level of suppression is given by the ratio Q_{Con}/Q_{Harr}, where Q_{Con} is the quality factor of the conventional clock, and Q_{Harr} is the lower pendulum quality factor used by Harrison.

From Eqs. (10.15) and (10.16) this reduces the indirect effect of the random external forces on the accumulated timing jitter variance by the ratio $\left(Q_{Con}/Q_{Harr}\right)^2$. Using the clock parameters shown in Fig. 10.5, $Q_{Harr} = 3000$ and $Q_{Con} = 6000$, Harrison's clock has approximately four times less indirect accumulated timing jitter than the conventional high-Q clock.

Harrison (1775:27) referred to the importance of the air's resistance and hence the use of low pendulum Q:

> of the Air's Resistance, want to be avoided, as many have foolishly imagined, but is of real or great Use.

Conclusions

It is not a surprise that Harrison's approach has been shown by the analysis in this paper to provide improved timekeeping performance over more conventional clock designs, given the excellent performance being achieved by a version of one of his pendulum clock designs (Betts, 2016). What is surprising is that his achievements have been underestimated and dismissed by so many for so long, a view reinforced by Harrison (1775:30) himself when he says:

> and therefore I may ask, if such a Matter be not highly worthy Encouragement, what other Sort of Ingenuity or Discovery in the World must be so? my Longitude Time-keeper, own Sister to this, excepted.

It is clear that the considerable effort Harrison spent in analysing, developing and optimising the geometry of his compensating mechanisms should have delivered far greater benefits to the wider understanding of high-precision clock design. It is hoped that this paper will help with both the wider recognition of his achievements and the wider adoption of his methods.

Annex A

In this annex we derive equations for the torques applied by the weighted composers in Harrison's grasshopper escapement.

Change in entry pallet torque with weighted composers

Figure 10.A1 shows the weighted composer resting on the entry pallet during the pendulum supplementary arc in zone 1.

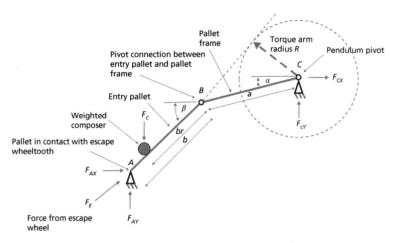

Fig. 10.A1 Entry pallet forces during supplementary arc. © David Harrison

The sum of horizontal and vertical forces caused by the weighted composer force F_C is zero, and

$$F_{AY} + F_{CY} - F_C = 0$$

$$F_{AX} + F_{CX} = 0. \tag{10.A1}$$

The sum of the horizontal and vertical forces acting on joints A and C is also zero, giving

$$F_{AY} = F_{AX} \tan \beta$$

$$F_{CY} = F_{CX} \tan \alpha. \tag{10.A2}$$

Finally, the sum of the turning moments acting about joint A is zero, giving

$$-F_C b(1-r)\cos\beta + F_{CY}(a\cos\alpha + b\cos\beta)$$
$$-F_{CX}(a\sin\alpha + b\sin\beta) = 0. \tag{10.A3}$$

Using Eqs. (10.A1), (10.A2) and (10.A3), the reaction force caused along the entry pallet arm by the weighted composer is given by

$$F_{AB} = \frac{F_C b(1-r)}{\tan\alpha(a\cos\alpha + b\cos\beta) - (a\sin\alpha + b\sin\beta)}. \tag{10.A4}$$

Using Eq. (10.A4) the total torque applied to the pendulum is then given by

$$\tau(\theta,\alpha,\beta)= F_E\left[1+\frac{\frac{F_C}{F_E}(1-r)}{\tan\alpha\left(\frac{a}{b}\cos\alpha+\cos\beta\right)-\left(\frac{a}{b}\sin\alpha+\sin\beta\right)}\right]R_0(1+4.5\theta), \quad (10.A5)$$

where R_0 is the torque arm radius at the pendulum's zero displacement angle.

Using a similar approach, the total torque applied to the pendulum when it is moving between the escaping arcs in zone 2 is found to be given by

$$\tau(\theta,\alpha,\beta)= F_E\left[1-\frac{\frac{F_C}{F_E}\left(1-r\frac{\cos\alpha}{\cos\beta}\right)}{\left(\tan\alpha\left(\frac{a}{b}\cos\alpha+\cos\beta\right)-\left(\frac{a}{b}\sin\alpha+\sin\beta\right)\right)}\right]R_0(1+4.5\theta). \quad (10.A6)$$

Using Harrison's drawing of the single pivot grasshopper, α reduces from $\sim 20°$ to $\sim 10°$, and β increases from $\sim 39°$ to $\sim 45°$, across the full pendulum running arc (Heskin, 2011). Using these changes in angle, the total change in escapement torque with arc of swing can be calculated using Eqs (10.A6) and (10.A7) for the entry pallet in zones 1 and 2.

Change in exit pallet torque with weighted composers

Figure 10.A2 shows a weighted composer resting on the exit pallet during the pendulum's supplementary arc. Using the same approach described in the previous section, the total escapement torque applied to the pendulum is found to be

$$\tau(\theta,\alpha,\beta)= F_E\left[1+\frac{\frac{F_C}{F_E}(1-r)}{\left(\tan\alpha\left(-\frac{a}{b}\cos\alpha+\cos\beta\right)+\left(\frac{a}{b}\sin\alpha+\sin\beta\right)\right)}\right]R_0(1+1.1\theta). \quad (10.A7)$$

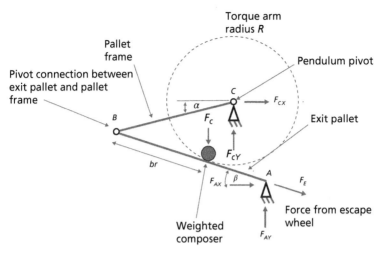

Fig. 10.A2 Exit pallet forces when pendulum is in its supplementary arc.
© David Harrison

Similarly, the total escapement torque applied to the pendulum when it is moving between the escaping angles in zone 2 is found to be given by

$$\tau(\theta,\alpha,\beta)=$$
$$F_E\left(1-\frac{\dfrac{F_C}{F_E}\left(1-r\dfrac{\cos\alpha}{\cos\beta}\right)}{\left(\tan\alpha\left(-\dfrac{a}{b}\cos\alpha+\cos\beta\right)+\left(\dfrac{a}{b}\sin\alpha+\sin\beta\right)\right)}\right)R_0\left(1+1.1\theta\right).$$
(10.A8)

From Harrison's drawing of the single pivot grasshopper, for the exit pallet, α increases from $\sim 10°$ to $\sim 20°$, and β reduces from $\sim 15°$ to $\sim 9°$ across the full pendulum arc of swing. These changes in angle can be used in Eqs. (10.A7) and (10.A8) to calculate the total change in escapement torque with pendulum arc for the exit pallet in zones 1 and 2.

Annex B

The change in rate of a pendulum clock caused by its escapement torque can be calculated using (Airy, 1830; Bogoliubov and Mitropolski, 1961)

$$\Delta\omega = -\frac{1}{2\pi\omega_0 A^3} \int_A^{-A} \frac{f(\theta)\theta}{\sqrt{1-\left(\dfrac{\theta}{A}\right)^2}}\, d\theta, \qquad (10.B1)$$

where $f(\theta)$ is the torque applied by the escapement, ω_0 is the natural frequency of the pendulum equal to π for Harrison's pendulum clock, A is the arc of swing of the pendulum and θ is the arc of the pendulum from its rest position.

Annex C

In this annex we derive an equation for effect of spring stiffness on the pendulum rate. Using the diagram in figure 10.C1, the deflection distance of the spring δ is given by

$$\delta = L_s \tan\theta, \qquad (10.C1)$$

where L_s is the length of the spring.

The force F_0 required at the end of the spring to cause this deflection is given by

$$\delta = \frac{F_0 L_s^3}{3EI}, \qquad (10.C2)$$

where E is the module of elasticity of the spring and I is its moment of inertia.

Using Eqs (10.C1) and (10.C2) we find the force at the end of the spring at a pendulum at arc θ to be

$$F_0 = \frac{3EI\tan\theta}{L_s^2}.$$

Finally, we find the torque applied to the pendulum by the force F_s in figure C1 to be

$$\tau_s \sim -\frac{3EI\sin\theta}{L_s} \sim -k\left(\theta - \frac{\theta^3}{6}\right), \qquad (10.C3)$$

where, $k = \dfrac{3EI}{L_s}$ and relates to the spring's stiffness. The spring's moment of inertia increases as the cube of its thickness. Hence, from Eq. (10.C3)

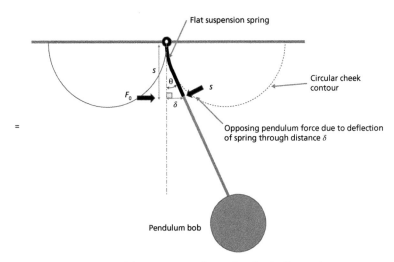

Fig. 10.C1 Flat spring deflecting around circular cheek. © David Harrison

the effective stiffness of the spring is greater when it is shorter and thicker and has a high module of elasticity.

Using Eq. (10.B1), this torque is found to cause a change in the pendulum frequency with running arc, given by

$$\Delta\omega \sim \frac{k}{\omega_0}\left(\frac{A^2}{16} - \frac{1}{2}\right). \tag{10.C4}$$

From this, a stiffer spring is seen to provide more compensation for the pendulum's circular error: $-\omega_0\dfrac{A^2}{16}$.

Acknowledgements

I would like to thank Mervyn Hobden and Martin Burgess for inspiring me to complete a PhD in nonlinear oscillator theory and providing helpful data and insights to the pursuing of the analysis set out in this paper. I would also like to thank Jonathan Betts for his helpful suggestions during the preparation of this paper.

References

Airy, G. B. (1830). On the disturbances of pendulums and balances and on the theory of escapements. *Transactions of the Cambridge Philosophical Society* III, Part I, 105–128.

Belleville, R. (2007). Exploring buoyancy. *Horological Science Newsletter*, Issue 2007–2 (April), 3–12.

Betts, J. (2016). Harrison's barometric compensation—keeping it simple: a description from practice. *Horological Science Newsletter*, Issue 2016–4 (September), 43–56.

Bogoliubov, N. N., and Mitropolski, Y. A. *Asymptotic Methods in the Theory of Non-Linear Oscillations*. Hindustan Publishing, 1961.

Demir, A., Mehrotra, A., and Roychowdhury, J. (2000). Phase noise in oscillators: a unifying theory and numerical methods for characterization. *IEEE Transactions on Circuits and Systems: Fundamental Theory and Applications*, 47(5), 655–674.

Haine, J. and Millington, A. (2016). Investigating the Harrison pendulum oscillator. *Horological Science Newsletter*, Issue 2016–3 (July), 2–18.

Harrison, J. (1775). *A Description Concerning Such Mechanism*. London, Jones, T. Available at https://ahsoc.contentfiles.net/media/assets/file/Concerning_Such_Mechanism.pdf. Accessed 12 December 2018.

Heskin, D. (2011). *Computer Aided Design of the Harrison Single Pivot Grasshopper Escapement Geometry*. Loughborough, UK: Soptera Publications.

Hoyng, P. (2014). Dynamics and performance of clock pendulums. *American Journal of Physics*, 82(11), 1053–1061.

Leeson, D. B. (1966). A simple model of feedback oscillator noise spectrum. *Proceedings of IEEE*, 54, 329–330.

Mazaheri, H. Hosseinzadeh, A. and Ahmadian, M. T. (2012). Nonlinear oscillation analysis of a pendulum wrapping on a cylinder. *Scientia Iranica*, 335–40.

Nelson, R. and Olsson, M. (1986). The pendulum—rich physics from a simple system. *American Journal of Physics*, 54(2), 112–121.

Schlichting, H. and Gersten, K. (2003). *Boundary-Layer Theory*. Berlin: Springer.

Woodward, P. (1997). Harrison's weighted escapement torque. *Horological Journal*, 196–197.

Update on Clock B

Rory McEvoy

Since its completion by Charles Frodsham & Co. in their Sussex workshop, Clock B has hung in only two locations. Firstly, as described in Chapter 7, in the horology workshop at the Royal Observatory, Greenwich where it had a desirably stable mounting on the stone column that supports the 18-ton Great Equatorial Telescope (Dolan, 2018). On 15 September 2016, Clock B was demounted and reinstalled in a gallery space on the first floor of what once served as living quarters for Observatory staff. Clock B was fastened to an external wall that carries a wooden floor, on which every year many thousands of visitors passed.

Over the first year, the clock was rated and left running in the gallery without further alteration. It was evidently affected by larger shifts in temperature (compared to those experienced in the workshop); however, the clock's performance was apparently unaffected by the vibrations caused by museum visitors, transmitted through the floor to the wall. For display purposes, the clock was reset every 2–3 months to keep its display to within a second of UTC.

In late 2017, computer monitoring equipment with connected GPS receiver was installed within the panelling around the clock's display case and a new adjustable temperature compensation tube was fashioned from CZ121 brass, following Martin Burgess's design. The two tubes are screwed together with a 0.25-mm thread pitch (or just over 100 threads per inch), to allow subtle increases of the height of the brass tube supporting the pendulum bob. The new device can be seen in Fig. Appendix 1, where the small gap in the middle defines the two separate parts—as the height of the tube is increased, this gap grows.

Plates 29–31 show screen captures from the Microset interface showing clock rate, temperature, and barometric pressure, taken across 2018. The first (saved May 25th) demonstrates that the clock was undercompensated when the new device was first fitted. The second (saved June 1st) shows that, after the first adjustment, the clock was overcompensated for temperature, proving the efficacy of the adjustable compensator. The third (saved August 31st) shows that adjustment of the tube height went too far and again rendered the clock undercompensated for temperature. Every adjustment of the telescopic tube requires careful rerating of the clock and this takes time and patience. Martin acquiesced that he should have incorporated Harrison's 'tin-whistle' type of

Figure Appendix 1 Clock B's pendulum bob, supported by the new adjustable temperature compensation. Credit: NMM.

adjustment during the making stage, thereby facilitating adjustment of the temperature compensation without major upset to the clock's rate. At the time of writing (December 2018), adjustment continues and is gradually progressing towards achieving the optimal temperature compensation.

As readers will no doubt be aware, the complexity of air density change through shifting temperature and air pressure will inevitably prohibit absolutely true timekeeping in a clock such as Clock B; however, its inherent stability has demonstrated that it is a reliable scientific instrument that can yield valuable data on the effects of subtle environmental changes to clocks running in free air and further our understanding of the physics of pendulum clocks.

At this juncture it seems fitting to thank two individuals: Don Saff, for identifying this opportunity to trial an almost forgotten theoretical approach to clockmaking. And finally, to Martin Burgess, who, with the help of many, has

> been able to prove beyond any doubt that...the remarkable claims he [Harrison] made for the accuracy of his regulators were not exaggerated. (Burgess, 1996)

References

Burgess, M. (1996). The scandalous neglect of Harrison's regulator science. In: Andrewes, W., ed., *The Quest for Longitude: The Proceedings of the Longitude Symposium, Harvard University, Cambridge Massachusetts, November 4–6, 1993*, pp:256–278. Harvard, Collection of Historical Scientific Instruments.

Dolan, G. (2018). The Great Equatorial Building (28-inch telescope dome). *The Royal Observatory Greenwich ... where east meets west*. Available at http://www.royalobservatorygreenwich. org/articles.php?article=919. Accessed 18 December 2018.

Index

Page numbers in italics are figures. Plates are shown by *pl*.